Tsunami - Damage Assessment and Medical Triage

Edited by Mohammad Mokhtari

Published in London, United Kingdom

IntechOpen

Supporting open minds since 2005

Tsunami – Damage Assessment and Medical Triage
http://dx.doi.org/10.5772/intechopen.82941
Edited by Mohammad Mokhtari

Contributors
Ghazala Naeem, Dong Li, Yunhua Zhang, Liting Liang, Jiefang Yang, Xun Wang, Tatiana Gouzeva, Diana Dimitrova, Mohammad Mokhtari, Jaime Santos-Reyes

Notice
Statements and opinions expressed in the chapters are these of the individual contributors and not necessarily those of the editors or publisher. No responsibility is accepted for the accuracy of information contained in the published chapters. The publisher assumes no responsibility for any damage or injury to persons or property arising out of the use of any materials, instructions, methods or ideas contained in the book.

First published in London, United Kingdom, 2021 by IntechOpen
IntechOpen is the global imprint of INTECHOPEN LIMITED, registered in England and Wales, registration number: 11086078, 5 Princes Gate Court, London, SW7 2QJ, United Kingdom
Printed in Croatia

British Library Cataloguing-in-Publication Data
A catalogue record for this book is available from the British Library

Additional hard and PDF copies can be obtained from orders@intechopen.com

Tsunami - Damage Assessment and Medical Triage
Edited by Mohammad Mokhtari
p. cm.
Print ISBN 978-1-83962-175-8
Online ISBN 978-1-83962-176-5
eBook (PDF) ISBN 978-1-83962-177-2

We are IntechOpen,
the world's leading publisher of
Open Access books
Built by scientists, for scientists

5,100+
Open access books available

126,000+
International authors and editors

145M+
Downloads

156
Countries delivered to

Our authors are among the

Top 1%
most cited scientists

12.2%
Contributors from top 500 universities

Interested in publishing with us?
Contact book.department@intechopen.com

Numbers displayed above are based on latest data collected.
For more information visit www.intechopen.com

Meet the editor

Mohammad Mokhtari was born in Nosara, Neyshabur and obtained his BSc, MSc, and Ph.D. from Azarabadegan, Southampton and Bergen Universities respectively. He also studied at Utrecht University and worked as a principal geophysicist at Norsk Hydro (Norway) and NIOC. As the Director of SRC at IIEES and Member of Board of Directors, he established the NBSN, was a Cofounder and Director of NCEP and a founding member of the Risk Management Excellence Center. He is also a member of the Passive Seismic Equipment Expert Panel, CTBTO, and a visiting researcher at GSA, Indian Ocean Tsunami Hazard Assessment. He was a member of the advisory board at IntechOpen publication. He is now chairman of ICG/NWIO-WG, a founding member of TERC and Director of Maga Makran project and VP of SPIRM Institute. He has supervised 32 MSc and 10 Ph.D. students. He has published over 75 papers, 100 conference presentations, and 7 books.

Contents

Preface

The history of tsunamis indicates that society is normally surprised and unprepared when it happens. It is always the history that enables us to plan for the future and to make it better and similarly, to apprehend the future losses from disasters like earthquakes and tsunamis. Studying past events using geological evidence gives us the best opportunity to plan for mitigation. Especially in the case of trans-oceanic tsunamis, which are rare, the only evidence we found earlier in many cases is historical documents. The past occurrence of such large tsunamis is usually preserved by geological features/structures along the coasts. The geological evidence not only gives insight into the past to understand what has happened, but it also helps us prepare for the future, including the fine-tuning of standard operating procedures for unexpected events and planning and designing of infrastructure for the development of a particular region. The long-term geological records provide opportunities to assess tsunami hazards more copiously. A more refined understanding of the long-term variations in time and the recurrence of giant tsunamis is essential for producing realistic vulnerability assessments for coastal communities. All these give us some guidelines to develop a technology that will help us with tsunami risk management and early warning systems. In addition to all of this, the local preparedness and education of the coastal communities is a very important aspect and requires worldwide attention.

In this book we have not fully covered the Probabilistic Tsunami Hazard Assessment (PTHA), but this methodology has been proven to be essential for tsunami risk reduction assessments. Using this method, we can prioritize high hazard locations for more detailed studies and identify the tsunami sources that contribute most to the hazard at a particular location. However, it is important to mention that tsunami risk management measures, such as planning evacuation routes would require a better understanding of high-resolution local-scale information.

The book has been divided into four sections. The first section is an introduction with the title of "Introductory Chapter: The Lessons Learned from Past Tsunamis and Todays Practice".

The second section covers "Monitoring of Tsunami/Earthquake Damages by Polarimetric Microwave Remote Sensing Technique", which is efficient and accurate monitoring and assessment for the fast response, management, and mitigation of the disasters. It is important to mention that Polarimetric remote sensing is an effective technique in the discrimination and recognition of ground objects. The section continues by considering the local tsunami on the Pakistan coast, where the earthquake and tsunami risk assessment and mitigation roadmap for Pakistan's coastal areas are being covered. The outbreak of devastating effects of an earthquake and its creation of health consequences, and medical providers of the population are discussed in the third section where proper medical triage and the provision of medical care in the outbreak of a traumatic disaster is a staged process. Section four will cover the risk of tsunamis in Mexico and the introduction of a preliminary tsunami early warning system.

This book is based on expert research with the objective of providing a collection of different aspects of tsunami characteristics. It is strongly believed that all the information provided here should play an important role in tsunami risk reduction and mitigation. The presented chapters have been peer-reviewed and accepted for publication. We would like to express our gratitude to the contributing authors who have been the key element in this achievement. Finally, we would like to express sincere gratitude to the IntechOpen publisher who initiated this book and guided and helped me with its completion.

Mohammad Mokhtari
Professor,
Chair of NWIO-WG at IOC/IGC UNESCO,
Australia

Founding Member of TERC and Technical Director
of Mega Project at Hormozgan University,
Iran

Section 1

Introduction

Introductory Chapter: The Lessons Learned from Past Tsunamis and Todays Practice

Mohammad Mokhtari

1. Introduction

It is always the history that gives us prospects to plan the future, based on past experiences, to make it better. Similarly, to apprehend the future losses from disasters like earthquakes and tsunamis, studying past events using geological evidence gives us the best opportunity to plan the mitigation. Especially in the case of trans-oceanic tsunamis, which are rare, the only evidence we found earlier in many cases is historical documents. The past occurrence of such large tsunamis is usually preserved by geological features/structures at coasts. The geological evidence not only gives insight into the past to understand what has had happened, but it also helps us prepare for the future, including the fine-tuning of standard operating procedures for unexpected events and planning and designing of infrastructure for the development of a particular region. The long-term geologic records provide opportunities to assess tsunami hazards more copiously. A more refined understanding of the long-term variations in time and the recurrence of giant tsunamis is essential for producing realistic vulnerability assessments for coastal communities.

The recent instrumental observations from geodesy and seismology together with the historical earthquake and tsunami data have a profound impact on the understanding of rupture patterns of large earthquakes and tsunami events. Nonetheless, the devastation caused by the 2004 Indian Ocean tsunami and the later major tsunamis made it clear that estimates of earthquake size and tsunami potential are woefully inadequate. Therefore, a more refined understanding of the long-term variations in timing and recurrence of giant tsunamis is essential for producing realistic vulnerability assessments for coastal communities. For example, if an earthquake similar to 1945 Makran occurs again, it will cause huge destruction at rapidly growing coastal cities of Iran, Pakistan, India, North of Oman sea, Yemen and UAE. During recent years, several studies in Oman, Pakistan, India, and Indonesia focused on North West Indian ocean tsunamis but Makran is one of the noted places which has deficient in data on tsunami size and frequency. No large-magnitude earthquake is known in the western Makran where the recorded seismicity is sparse. By contrast, large-magnitude and frequent earthquakes characterize the eastern Makran. This geographical dissimilarity in seismicity is attributed to a hypothetical segmentation of the subduction zone or to a locked plate boundary that experiences great earthquakes with long repeat times in the west [1].

2. Tsunami sources

Knowledge of tsunami sources is important in tsunami warning and mitigation. Where tsunami detectors should be deployed and maintained? What incoming tsunami heights should modelers assume in making inundation maps? Which coasts have the greatest need of mitigation measures? The answers depend on estimates of tsunami size and frequency, which in turn may vary from one tsunami source to the next.

Let's now look at the case of the Indian Ocean. There are two major tsunamigenic sources in the Indian Ocean region. One is the Andaman-Sumatra subduction zone and the other one is the Makran subduction zone. The results of the studies show that the Andaman-Sumatra subduction zone being affected by many major tsunamis in the past. However, the Makran subduction zone, which located between Iran and Pakistan to the north of the Arabian Sea, neither there is many historical records nor any long-term studies. Though recent studies established the fact that Oman, Pakistan, Iran, and India were repeatedly hit by tsunamis in historical times, it is still unclear how strong were these events and what was the height of the tsunami waves. For example, the investigation found that boulder deposits and fine-grained sediments from the north coast of Oman are evidence for much larger pre-historic tsunami events in the region. It is interesting to note that the seismicity differs in the eastern and western parts of the Makran subduction zone, with a boundary at about the Iran/Pakistan border. No large-magnitude earthquake is known in the western Makran where the recorded seismicity is sparse, raising the question of locked or aseismic? By contrast, large-magnitude and more frequent earthquakes characterize the eastern Makran. However, the evidence found from the Oman coast is in contrast to the theory of division of the Makran subduction zone in two. Therefore, the earthquake might have at least partially ruptured the western Makran, which would imply that the western Makran is not completely unlocked. Consequently, the hazard scenarios prepared based on historical data underestimate the tsunami threat in the Northern Arabian Sea. Keeping in view the indistinct structure of the Makran subduction zone as well as inconsistent historical tsunami records. To help the situation a Paleo-tsunami studies can play an important role in further assessment of the region. In this case the dated deposits can allow us to estimate the times and recurrence intervals of past tsunamis. The obtained result can guide the mitigation efforts and may reduce major losses due to future tsunamis. This study now being initiated and it will happen in a regional context which would reduce the uncertainty where a repeated location would be investigated. Parallel to above in the Makran region the onshore part is also important for better understanding the structural framework of the area, so the following work has been conducted and shown the result is unique.

3. Recent active seismic data

To improve a better understanding of the Makran subduction zone structural model recently, three active seismic profiles (**Figure 1**) have been acquired onshore west Makran. Each profile was about 200 km long, with a shot-point interval of 20 km and the receiver interval of 700 m [2]. **Figure 2** shows the results of the model analysis. The high signal to noise ratio and resolution is exclusively valuable, which give an exclusive evidence on layering of the accretionary margin and also the oceanic Moho with rather high resolution. The offshore continuation of these

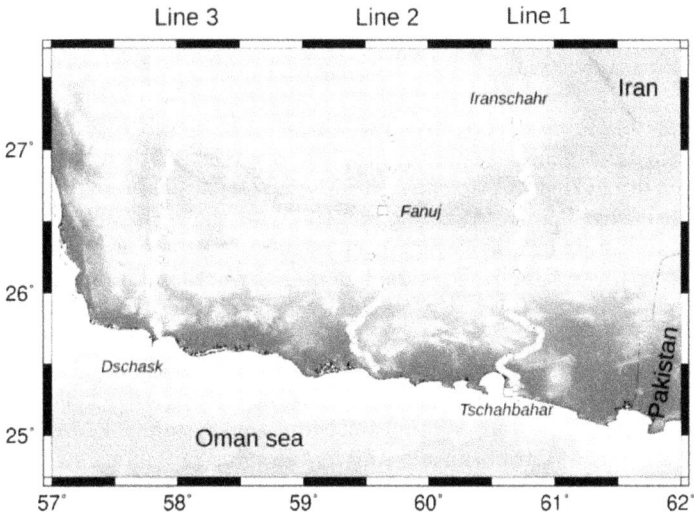

Figure 1.
Location of onshore active seismic data in the West Makran. The yellow lines indicate the lines. The open circle on the lines indicate the shot point locations.

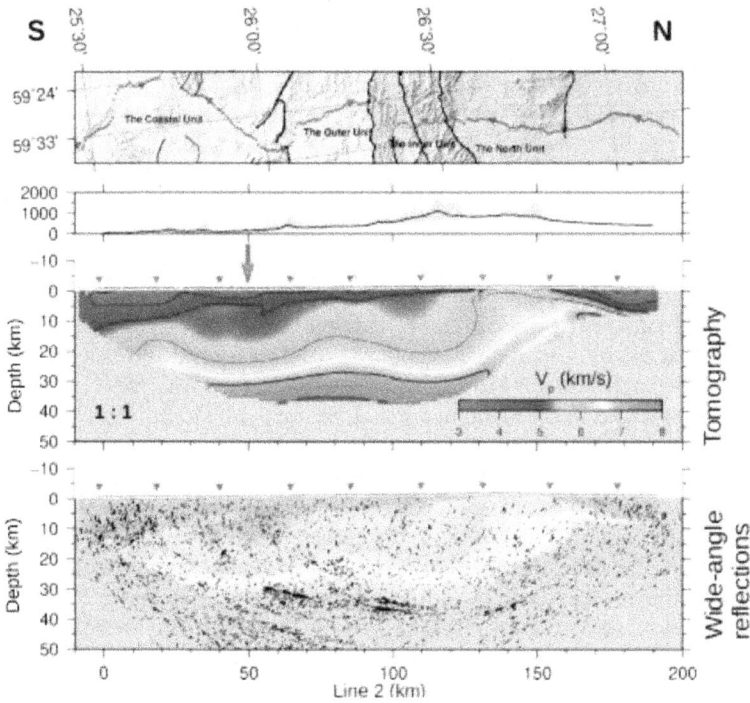

Figure 2.
Results for seismic profile (Line 2). P-wave velocity (Vp) of tomographic inversions are color coded; unresolved regions are clipped; contours are Vp with labels in km/s. Shot locations are indicated by red inverted triangles. Line-drawing elements (black dots, point out reflective bands) migrated with tomographic Vp model are overlaid on Vp model. Tectono-structural units of Makran (M) is indicated on top of panel.

lines is ongoing and hope soon we will cover with high resolution the entire wester Makran margin. These will provide the major parameters of future Makran megathrust fault.

4. Tsunami risk reduction measures

4.1 Early warning system

Following the Sanriku Tsunami in 1933 the first tsunami early warning system was established in Japan. Tsunami warning center was established in 1946 after Unimak tsunami in Hawaii. After the 1964 Mw 9.2 Great Alaska earthquake and Tsunami the Pacific Tsunami Warning Centre, and in the 1960 Valdivia Mw 9.3–9.6 earthquake and tsunami in Chile have been established [3]. Tsunami early warning systems had not been established in the Indian Ocean at the time of the 2004 Indian Ocean Tsunami. Again, it was only following the major disaster that the regional tsunami early warning systems were established in the Indian Ocean. Recently Indian Ocean Tsunami Warning and Mitigation System, IOTWS; the North-eastern Atlantic and the Mediterranean Sea (North-eastern Atlantic, the Mediterranean and Connected Seas Tsunami Warning and Mitigation System, NEAMTWS), and the Caribbean region (Caribbean and Adjacent Regions Early Warning System, CARIBE-EWS) has been established. Additionally, national tsunami early warning systems were established in the Indian Ocean (e.g., Indonesian Tsunami Early Warning Systems, Ina-TEWS) [4].

In the Makran region using different earthquakes and tsunami, the estimated time for tsunami waves to hit the coastlines of Iran and Pakistan and Oman is between 15 and 20 minutes. This strongly suggests a major need for the establishment of a tsunami early warning center in the region. It should be mentioned at the present there two national early warning systems one in India and the other one in Oman. It is strongly necessary that these centers to be task as waring dissemination in the region as well till a new center with task region being established for the here we should add 4.

4.2 PTHA modelling

As there are different causes for tsunami generation (earthquakes, landslides, volcanic activity, meteorological events, and asteroid impacts) number of geophysical assessments should be in principle conducted. It is important to mention one the mythology that can be used in understanding a tsunami hazard and risk reduction measures is the application of Probabilistic Tsunami Hazard Analyses (PTHAs) among other methods in the world in a global, regional, and local scales. Tsunami hazard assessment methodologies are not standardized and recent destructive tsunamis It is difficult to quantify the hazard of Hazard assessments need to consider how the tsunami hazard information will be used, the relevant tsunami sources, the propagation, and inundation models, and whether the assessment is probabilistic or scenario-based. The determination of maximum earthquake magnitude is difficult. For example, the magnitude range for the Sunda Arc is 9.0–9.6 whereas the range for the Makran subduction zone is 8.1–9.3. Predictions of tsunami inundation are influenced by earthquake characteristics, models, bathymetric and topographic data resolution, land cover roughness, and tides. Tuning models to historical events can increase accuracy and return periods can be calculated from pre-historic tsunamis. Can help modelers and end-users to agree on methodologies and how to best deal with uncertainties. Before the occurrence of the 2004 Indian Ocean Tsunami, only a few PTHA's were carried out. However, the PTHA method gained momentum due to the many tsunami hazard assessments which included the deterministic hazard studies; to inform the risk reduction to the stakeholder and government authorities. The development of modern PTHA techniques being done by Geist and Parsons [5], and some-more studies using the methodology followed later.

The unexpected measure of the disaster caused by the 11 March 2011, Tohoku earthquake and tsunami in Japan in a country that has invested much effort on tsunami preparation, again showed that our understanding of tsunami hazard and risk is being been limited. In essence, this event highlighted the need for incorporating more complex phenomena such as variable slip and quantification of source uncertainties in PTHA analysis.

5. Conclusion and future work

The recent observations from geodesy and seismology in combination with historical tsunami data have a profound impact on the understanding of rupture patterns of large earthquakes and its consequent tsunami events. However, the 2004 Indian Ocean tsunami which caused a devastating effect made it clear that estimates of earthquake size and tsunami potential are woefully inadequate. So, some powerful methodology requires to reduce this short come. Among this which recently has gained momentum is the Probabilistic Tsunami Hazard Assessment or simply PTHA. It should be noted that many aspects of PTHA need to be revised for the future application of this methodology. Despite a major advance in the PTHA methods, the standards on how to conduct PTHA assessments is lacking now and in addition, the probabilistic risk assessments need to be revisited. The lack of systematic data for building the models for earthquake sources for example shallow megathrust zone close to the trench. It should also be noticing the inclusion of landslide and splay faulting [6] as other hazard strengthening elements due to the lack of an instrumental record of occurrence has proven. [7, 8] have noticed that tsunami risk combines the calculations of PTHA, exposure, and fragility. So, the tsunami risk calculations will include both human populations and the main infrastructure. Thus, the development of future probabilistic risk assessments will rely critically on developing tsunami fragility curves and systematic standards for PTHA calculations. The use of the logic tree framework in PTHA is gaining momentum. It should be noted due to the ease of technical implementation of the logic trees being used frequently. It is a powerful tool to organize the way of thinking in situations where alternative models, in which the analysts have different degrees of confidence, might apply. The end product can greatly enhance tsunami risk reduction efforts. Finally, it is important to mention Paleo tsunami researches is a powerful tool that can lead to a better constraining Mmax and the rate of large tsunami events.

Author details

Mohammad Mokhtari[1,2]

1 IOC/IGC UNESCO, Australia

2 TERC at The University of Hormozgan, Iran

*Address all correspondence to: m7mokhtari@gmail.com

IntechOpen

References

[1] Mokhtari et al. (2019). A review of the seismotectonics of the Makran Subduction Zone as a baseline for Tsunami Hazard Assessments. Geosci. Lett. DOI: 10.1186/s40562-019-0143-1

[2] Haberland C, et al. (2020). Anatomy of a crustal-scale accretionary complex: Insights from deep seismic sounding of the onshore western Makran subduction zone, Iran: Geology, v. 48, DOI: 10.1130/G47700.1

[3] Abe K. (1979). Size of great earthquake of 1837-1974 inferred from tsunami data. Journal of Geophysical Research. **84**:1561-1568

[4] Lauterjung et al. (2010). The challenge of installing a tsunami early warning system in the vicinity of the Sunda Arc, Indonesia. Natural Hazards and Earth System Sciences. **10**(4). DOI: 10.5194/nhess-10-641-2010

[5] Geist EL, Parsons T. (2006). Probabilistic Analysis of Tsunami Hazards. Natural Hazards. *37*:277-314. DOI: 10.1007/s11069-005-4646-z

[6] Mokhtari M. (2014). The role of splay faulting in increasing the devastation effect of tsunami hazard in Makran, Oman Sea. Arabian Journal of Geosciences. **8**. DOI: 10.1007/s12517-014-1375-1

[7] Løvholt F, Kuhn D, Bungum H, Harbitz C, Glimsdal S. (2012). Historical tsunamis and present tsunami hazard in eastern Indonesia and southern Philippines. Journal of Geophysical Research. **117**. DOI: 10.1029/2012JB009425

[8] Løvholt F, Glimsdal S, Smebye H, Griffin J, Davies G. (2015). Tsunami Methodology and Result Overview. Technical Report, UN-ISDR Global Assessment Report. Geoscience Australia: Canberra. p. 2015

Section 2

Tsunami Damage Estimate

Monitoring of Tsunami/ Earthquake Damages by Polarimetric Microwave Remote Sensing Technique

Dong Li, Yunhua Zhang, Liting Liang, Jiefang Yang and Xun Wang

Abstract

Polarization characterizes the vector state of EM wave. When interacting with polarized wave, rough natural surface often induces dominant surface scattering; building also presents dominant double-bounce scattering. Tsunami/earthquake causes serious destruction just by inundating the land surface and destroying the building. By analyzing the change of surface and double-bounce scattering before and after disaster, we can achieve a monitoring of damages. This constitutes one basic principle of polarimetric microwave remote sensing of tsunami/earthquake. The extraction of surface and double-bounce scattering from coherency matrix is achieved by model-based decomposition. The general four-component scattering power decomposition with unitary transformation (G4U) has been widely used in the remote sensing of tsunami/earthquake to identify surface and double-bounce scattering because it can adaptively enhance surface or double-bounce scattering. Nonetheless, the strict derivation in this chapter conveys that G4U cannot always strengthen the double-bounce scattering in urban area nor strengthen the surface scattering in water or land area unless we adaptively combine G4U and its duality for an extended G4U (EG4U). Experiment on the ALOS-PALSAR datasets of 2011 great Tohoku tsunami/earthquake demonstrates not only the outperformance of EG4U but also the effectiveness of polarimetric remote sensing in the qualitative monitoring and quantitative evaluation of tsunami/earthquake damages.

Keywords: disaster monitoring, damage evaluation, tsunami, earthquake, microwave remote sensing, synthetic aperture radar (SAR), polarimetric SAR (PolSAR), polarimetric decomposition, scattering model, unitary transformation

1. Introduction

Tsunami and earthquake seriously endanger people's lives and properties. Efficient and accurate monitoring and assessment are of crucial importance for the fast response, management, and mitigation of the disasters [1–3]. Compared with the optical remote sensing, microwave remote sensing technology such as synthetic aperture radar (SAR) has been widely applied to monitoring natural and human-induced disasters for its all-day and all-weather working capacity [4].

Polarization is an essential property of the electromagnetic wave [5–8]. The polarization state of wave will change when interacting with ground object. For example, rough natural surface such as land and water often induces the strong Bragg surface scattering, while building often presents the dominant double-bounce scattering because of the dihedral corner reflectors formed by ground and the vertical wall of building. Therefore, by analyzing the polarization of the scattering wave, we can acquire the physical and geometrical information regarding the object. This is the main task of SAR polarimetry (PolSAR) [9–11].

Tsunami is often accompanied by earthquake and flooding [1–3]. It damages and inundates the buildings and causes the collapse of the ground-wall dihedral structures as well as the enhancement of the direct surface scatterers. Therefore, by analyzing the power of double-bounce scattering and surface scattering before and after the event, we can achieve an efficient monitoring of the disasters. This simple strategy has been successfully adopted in the polarimetric microwave remote sensing of tsunami/earthquake [12–21].

Nonetheless, the extraction of double-bounce scattering and surface scattering from PolSAR image is not so straightforward because each pixel in PolSAR is a 3×3 complex coherency matrix $\langle[T]\rangle$ with nine degrees of freedom (DoF). A widely used approach to achieve this is to decompose $\langle[T]\rangle$ on the canonical scattering models [22]. The first such decomposition was devised by Freeman and Durden [23] which expands $\langle[T]\rangle$ on the surface scattering, double-bounce scattering, and volume scattering (describes the complex scattering in vegetation area). This three-component decomposition, however, is responsible for only five DoF of $\langle[T]\rangle$ because of the symmetric reflection assumption. This assumption was tackled by Yamaguchi et al. [24] by introducing a fourth helix component and two additional models of volume scattering. The resulted four-component decomposition (Y4O) then only leaves three DoF unaccounted: the $(1, 3)$ element of $\langle[T]\rangle$, i.e., T_{13}, and the real part of the $(2, 3)$ element of $\langle[T]\rangle$, i.e., $\mathrm{Re}\{T_{23}\}$. A same target will present differently by a simple rotation about the line of sight of radar. Deorientation should be first conducted on $\langle[T]\rangle$ to eliminate the influence [25]. As a result, $\mathrm{Re}\{T_{23}\}$ changes to zero and Y4O with rotation (Y4R) accounts for seven DoF [26]. Based on Y4R, Sato et al. [27] further proposed to add a new model to characterize volume scattering generated by even-bounce structure. However, Sato's extended Y4R (S4R) still leaves T_{13} unaccounted. To solve this, Singh et al. [28] in 2013 proposed a general four-component decomposition (G4U) based on a special unitary matrix. G4U enables T_{13} included in the accounted models by conducting unitary transformation to the rotated version of $\langle[T]\rangle$. Singh et al. [28] claimed that G4U could make full use of polarimetric parameters. As a result, in comparison with the four-component decompositions such as S4R and Y4R, G4U could enhance double-bounce scattering power over urban area while enhancing surface scattering contribution over an area where surface scattering is preferable [28]. This makes G4U very suitable to the remote sensing of tsunami/earthquake [16, 20] and establishes G4U the state-of-the-art four-component scattering power decomposition [29, 30].

This chapter is dedicated to enable an extension to G4U for better monitoring of tsunami/earthquake disaster. It is indicated that the unitary transformation in G4U adds a T_{13}-related but redundant balance equation to the original self-contained equation system in Y4R and S4R. Then T_{13} is accounted for by G4U, but we obtain no exact solution to the system but the approximate ones. The general expression of the approximate solutions enables a generalized G4U (GG4U), while G4U and S4R represent two special forms. A dual G4U (DG4U) is also obtained. The general solution indicates that G4U cannot always enhance the double-bounce scattering power over urban area nor strengthen the surface scattering power over the area where surface scattering is dominant unless we adaptively integrate G4U and DG4U for an extended G4U (EG4U). Experiments on the PolSAR images of Miyagi Prefecture, Japan, acquired by the L-band spaceborne ALOS-PALSAR system before and after the March 11, 2011, Off-Tohoku 9.0 tsunami/earthquake demonstrate not only the outperformance of EG4U but also the effectiveness of polarimetric remote sensing in the monitoring of tsunami/earthquake disaster.

The remainder of this chapter is arranged as follows. Section 2 presents the basic principle of PolSAR and the polarization descriptors first. The advanced four-component scattering power decompositions are then described in Section 3 to develop the EG4U. By decomposing the ALOS-PALSAR datasets of the 2011 great Tohoku tsunami/earthquake using EG4U, Section 4 evaluates and analyzes the polarimetric monitoring of disaster damages further. The chapter is eventually concluded in Section 5.

2. SAR polarimetry and polarization descriptors

SAR is an active microwave remote sensing technique dedicated to acquire the large-scaled 2D coherent image of the earth's surface reflectivity [9]. It transmits microwave pulses and receives the backscattering from the illuminated terrain to synthesize a high spatial resolution image. Such an active operation enables SAR an all-day working capacity independent of solar illumination. In addition, operating in the microwave region of electromagnetic spectrum avoids the effects of rain and clouds, which allows SAR an almost all-weather continuous monitoring of the earth surface [9].

Polarization characterizes the vector state of the electromagnetic wave. The polarization state of wave will change when interacting with a ground object. By processing and analyzing such change of polarization, we can obtain the material, roughness, shape, and orientation information regarding the object. The core of this change is the (Sinclair) scattering matrix $[S]$ of the object, which transforms the incident electric filed E^I into the scattered electric filed E^S [31]:

$$E^S = \frac{e^{-jkr}}{r}[S]E^I \rightarrow \begin{bmatrix} E^S_H \\ E^S_V \end{bmatrix} = \frac{e^{-jkr}}{r} \begin{bmatrix} S_{HH} & S_{HV} \\ S_{VH} & S_{VV} \end{bmatrix} \begin{bmatrix} E^I_H \\ E^I_V \end{bmatrix} \tag{1}$$

where r denotes the distance from radar to ground object, k is the wave number, and subscript H or V represents the horizontal or vertical polarization. Matrix $[S]$ is obtained by first transmitting H-polarized wave (E^I_H) and receiving scatterings in H- and V-polarization simultaneously to measure the first column S_{HH} and S_{VH} and then transmitting V-polarized wave (E^I_V) and also receiving in H- and V-polarization simultaneously for the second column S_{HV} and S_{VV}. In reciprocal backscattering, we have $S_{HV} = S_{VH}$ and matrix $[S]$ covers five DoF then.

Generally, almost all the ground scatterers are situated in the dynamically changing environment and subjected to spatial and/or temporal variations [32]. Such scatterer is called the distributed target, and we can no longer model its scattering with a determined scattering matrix $[S]$. The 3×3 coherency matrix $\langle[T]\rangle$ is then constructed as the statistical average of the acquired scatterings to describe the second-order moment of the fluctuations [9]:

$$\langle[T]\rangle = \langle \boldsymbol{kk}^\dagger \rangle = \begin{bmatrix} T_{11} & T_{12} & T_{13} \\ T_{21} & T_{22} & T_{23} \\ T_{31} & T_{32} & T_{33} \end{bmatrix}, \boldsymbol{k} = \frac{1}{\sqrt{2}} \begin{bmatrix} S_{HH} + S_{VV} \\ S_{HH} - S_{VV} \\ 2S_{HV} \end{bmatrix} \tag{2}$$

where $\langle \cdot \rangle$ and superscript † represent the operations of ensemble average and conjugate transpose and k denotes the Pauli vector. The spatial/temporal depolarization pushes the DoF of $\langle [T] \rangle$ to nine. Therefore, different from the conventional SAR image, each pixel in PolSAR image is not a complex number but a 3×3 coherency matrix $\langle [T] \rangle$.

The coherency matrix $\langle [T] \rangle$ in Eq. (2) is expressed in the H-V polarization basis; we can also formulate it in some other orthonormal basis by simply taking the unitary transformation of $\langle [T] \rangle$:

$$\text{Unitary}(\langle [T] \rangle) \overset{\text{def}}{=} [U_3] \langle [T] \rangle [U_3]^\dagger \tag{3}$$

where $[U_3]$ is the special unitary matrix that describes the relationship between H-V and the new orthonormal basis. Target deorientation is just based on the real rotation matrix [25]:

$$[U_3(\theta)] = \begin{bmatrix} 1 & 0 & 0 \\ 0 & \cos 2\theta & \sin 2\theta \\ 0 & -\sin 2\theta & \cos 2\theta \end{bmatrix}, 2\theta = \frac{1}{2} \tan^{-1} \left(\frac{2 \operatorname{Re} \{T_{23}\}}{T_{22} - T_{33}} \right) \tag{4}$$

Combining Eq. (4) into Eq. (3), the deoriented coherency matrix $\langle [T'] \rangle$ is

$$\langle [T'] \rangle = [U_3(\theta)] \langle [T] \rangle [U_3(\theta)]^\dagger = \begin{bmatrix} T'_{11} & T'_{12} & T'_{13} \\ T'_{21} & T'_{22} & j\operatorname{Im}\{T'_{23}\} \\ T'_{31} & j\operatorname{Im}\{T'_{32}\} & T'_{33} \end{bmatrix}. \tag{5}$$

Deorientation makes T'_{23} become purely imaginary and reduces DoF from nine to eight. In order to eliminate the imaginary part further, Singh et al. developed an imaginary rotation matrix [28]:

$$[U_3(\varphi)] = \begin{bmatrix} 1 & 0 & 0 \\ 0 & \cos 2\varphi & j\sin 2\varphi \\ 0 & j\sin 2\varphi & \cos 2\varphi \end{bmatrix}, 2\varphi = \frac{1}{2} \tan^{-1} \left(\frac{2\operatorname{Im}\{T'_{23}\}}{T'_{22} - T'_{33}} \right). \tag{6}$$

A coherency matrix $\langle [T''] \rangle$ with zero T''_{23} is then achieved:

$$\langle [T''] \rangle = [U_3(\varphi)] \langle [T'] \rangle [U_3(\varphi)]^\dagger = \begin{bmatrix} T''_{11} & T''_{12} & T''_{13} \\ T''_{21} & T''_{22} & 0 \\ T''_{31} & 0 & T''_{33} \end{bmatrix}. \tag{7}$$

3. Advanced four-component scattering power decompositions

Polarimetric incoherent decomposition plays an important role in the discrimination and recognition of the distributed target [22]. It pursues the scattering mechanism of the unknown target by extracting the dominant or average target (such as the Huynen-type phenomenological dichotomies [7, 32] and the eigenvalue/eigenvector-based target decompositions [9, 33]) from $\langle [T] \rangle$ or expanding $\langle [T] \rangle$ on the canonical models (such as the model-based target decompositions [23–28]). Among these decompositions, the four-component scattering power decompositions such as Y4R, S4R, and G4U have been a hot topic recently [29].

3.1 Y4R and S4R

Y4R and S4R decompose the target by linearly expanding matrix $\langle[T']\rangle$ on the four canonical scattering models, as illustrated in **Figure 1**:

$$\langle[T']\rangle = f_S\langle[T'_S]\rangle + f_D\langle[T'_D]\rangle + f_V\langle[T'_V]\rangle + f_C\langle[T'_C]\rangle \tag{8}$$

where $\langle[T'_S]\rangle$, $\langle[T'_D]\rangle$, $\langle[T'_V]\rangle$, and $\langle[T'_C]\rangle$ denote the surface scattering model, the double-bounce scattering model, the volume scattering model, and the helix scattering model, respectively:

$$\langle[T'_S]\rangle = \begin{bmatrix} 1 & \beta & 0 \\ \beta^* & |\beta|^2 & 0 \\ 0 & 0 & 0 \end{bmatrix}, \langle[T'_D]\rangle = \begin{bmatrix} |\alpha|^2 & \alpha & 0 \\ \alpha^* & 1 & 0 \\ 0 & 0 & 0 \end{bmatrix},$$

$$\langle[T'_V]\rangle = \begin{bmatrix} a & d & 0 \\ d & b & 0 \\ 0 & 0 & c \end{bmatrix}, \langle[T'_C]\rangle = \frac{1}{2}\begin{bmatrix} 0 & 0 & 0 \\ 0 & 1 & \pm j \\ 0 & \mp j & 1 \end{bmatrix}. \tag{9}$$

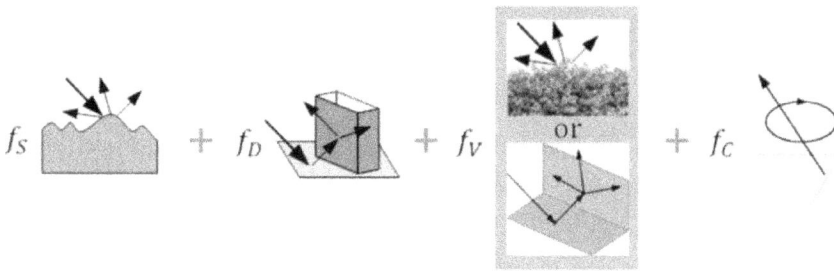

Figure 1.
The canonical models involved in the four-component model-based scattering power decompositions.

Parameters f_S, f_D, f_V, and f_C in Eq. (8) represent the contributions of the four components; β and α in $\langle[T'_S]\rangle$ and $\langle[T'_D]\rangle$ are complex parameters; $a, b, c,$ and d in $\langle[T'_V]\rangle$ are real constants satisfying $a + b + c = 1$, which involve in four volume scattering models and are adaptively selected according to the branch conditions [27, 28]. Combining Eqs. (8) and (9), the S4R/Y4R scattering balance equation system on unknowns $f_S, f_D, f_V, f_C, \alpha,$ and β is formulated [26, 27]:

$$\begin{cases} f_S + f_D|\alpha|^2 + f_V a = T'_{11} & -1) \\ f_S\beta + f_D\alpha + f_V d = T'_{12} & -2) \\ f_S|\beta|^2 + f_D + f_V b + \dfrac{f_C}{2} = T'_{22} & -3) \\ \pm j\dfrac{f_C}{2} = j\mathrm{Im}\{T'_{23}\} & -4) \\ f_V c + \dfrac{f_C}{2} = T'_{33} & -5) \end{cases} \tag{10}$$

Nevertheless, we obtain no scattering balance equation on T'_{13} in Eq. (10). Hence, there always exists a T'_{13} -related unaccounted residue in Y4R and S4R.

3.2 G4U

To model T'_{13}, G4U uses $[U_3(\varphi)]$ to conduct unitary transformation to both sides of Eq. (10) first and then eliminates the influence of φ [28]. As a result, an additional balance equation is brought into G4U, and we obtain the following scattering balance equation system [30]:

$$
\begin{cases}
f_S + f_D|\alpha|^2 + f_V a = T'_{11} & -1) \\
f_S\beta + f_D\alpha + f_V d = T'_{12} + T'_{13} \\
f_S\beta + f_D\alpha + f_V d = T'_{12} - T'_{13}
\end{cases} \Big\} -2) \\
\begin{cases}
f_S|\beta|^2 + f_D + f_V b + \dfrac{f_C}{2} = T'_{22} & -3) \\
\pm j\dfrac{f_C}{2} = j\mathrm{Im}\{T'_{23}\} & -4) \\
f_V c + \dfrac{f_C}{2} = T'_{33} & -5)
\end{cases}
\tag{11}
$$

Comparing Eq. (11) with Eq. (10), we can find that Eq. (11–2) gives a dichotomy to Eq. (10–2). The redundancy makes Eq. (11) have no such exact solution like Eq. (10) but some approximate ones. In G4U, Singh et al. preferred the first equation of (11–2) only.

3.3 GG4U: generalization of G4U

Obviously, Eq. (11) provides us a generalized G4U (GG4U). Here we focus on the general solution to (11) for the unknowns $f_S, f_D, f_V, f_C, \alpha$, and β. Let

$$
\begin{cases}
S = T'_{11} - f_V a \\
C_1 = T'_{12} + T'_{13} - f_V d \\
C_2 = T'_{12} - T'_{13} - f_V d \\
D = T'_{22} - f_V b - \dfrac{f_C}{2} \\
C = \dfrac{1+\mu}{2}C_1 + \dfrac{1-\mu}{2}C_2
\end{cases}
\tag{12}
$$

where μ is a real constant. Then Eq. (11) can be rearranged as

$$
\begin{cases}
f_S + f_D|\alpha|^2 = S & -1) \\
f_S\beta + f_D\alpha = C & -2) \\
f_S|\beta|^2 + f_D = D & -3) \\
f_C = 2|\mathrm{Im}\{T'_{23}\}| & -4) \\
f_V = \dfrac{1}{2c}(2T'_{33} - f_C) & -5)
\end{cases}
\tag{13}
$$

Eq. (13) comprises of five equations and six unknowns. Following Freeman-Durden [23] and Yamaguchi et al. [24], we can fix α or β in terms of the sign of $S - D$ for the superior between surface scattering and double-bounce scattering:

$$\begin{cases} BC > 0 \Rightarrow \text{dominant surface scattering} \Rightarrow \alpha = 0 \\ BC \leq 0 \Rightarrow \text{dominant double} - \text{bounce scattering} \Rightarrow \beta = 0 \end{cases} \tag{14}$$

where $BC = S - D$. Combining Eqs. (13) and (14), we can then simply obtain the scattering power of each of the four components, i.e., the surface scattering power P_S, the double-bounce scattering power P_D, the volume scattering power P_V, and the helix scattering power P_C:

$$\begin{cases} P_S = f_S\left(1 + |\beta|^2\right) = \begin{cases} S + \dfrac{|C|^2}{S}, BC > 0 \\[3mm] S - \dfrac{|C|^2}{D}, BC \leq 0 \end{cases} \\[10mm] P_D = f_D\left(1 + |\alpha|^2\right) = \begin{cases} D - \dfrac{|C|^2}{S}, BC > 0 \\[3mm] D + \dfrac{|C|^2}{D}, BC \leq 0 \end{cases} \\[10mm] P_C = f_C H\left(T'_{33} - |\text{Im}\{T'_{23}\}|\right) \\[3mm] P_V = \dfrac{1}{2c}\left(2T'_{33} - P_C\right) \end{cases} \tag{15}$$

where $H(\cdot)$ denotes the Heaviside step function, which is used here to adjust the value of P_C for nonnegative P_V ruling [27]. It can be easily validated that $P_S + P_D + P_V + P_C = T'_{11} + T'_{22} + T'_{33}$. Thus GG4U gives a decomposition of scattering power.

3.4 Special decompositions

By taking appropriate value to μ, we can have some different decompositions, which are denoted as $\mathcal{G}(\mu)$. Here we are particularly interested to the following special cases of $\mathcal{G}(\mu)$.

Case (1): $\mathcal{G}(+1) := $ G4U

$$C = C_1 = T'_{12} + T'_{13} - f_V d = C_{\text{G4U}}. \tag{16}$$

This is just the parameter C used in G4U. GG4U changes to G4U in this case.

Case (2): $\mathcal{G}(-1) := $ DG4U

$$C = C_2 = T'_{12} - T'_{13} - f_V d. \tag{17}$$

This acts as the complement of case (1); thus we name it the dual G4U (DG4U).

Case (3): $\mathcal{G}(0) := $ S4R

$$C = \frac{C_1 + C_2}{2} = T'_{12} - f_V d = C_{\text{S4R}}. \tag{18}$$

This is the parameter C used in S4R, i.e., S4R also shows a special form of GG4U. Hence, the essential difference between S4R and G4U just lies in the different definition of parameter C in Eqs. (16) and (18). The unitary transformation is just to enable the T'_{13} entry contained in C_{G4U} and finally in P_S and P_D. Parameter C defined in Eq. (12) is a generalization of C_{G4U} and C_{S4R}.

3.5 Theoretical evaluation of S4R and G4U

S4R can improve Y4R by strengthening the double-bounce scattering in urban area [27]. Singh et al. [28] indicated that G4U could further improve S4R in this aspect by strengthening surface scattering in the area where surface scattering is preferable to double-bounce scattering, while increasing the double-bounce scattering in the urban area where the double-bounce scattering is preferable to surface scattering. By combining the ruling in Eq. (14), we can formulate these observations as

$$\begin{cases} P_S^{G4U} \geq P_S^{S4R}, BC > 0 \\ P_D^{G4U} \geq P_D^{S4R}, BC \leq 0 \end{cases}. \tag{19}$$

In terms of the general expression of P_S and P_D in (15), here we give a simple validation to Eq. (19) by combining $\mu = 0$ and $\mu = 1$ into Eqs. (12) and (15):

$$\begin{cases} P_S^{G4U} = S + \dfrac{|C_1|^2}{S} \\ P_S^{S4R} = S + \dfrac{|C_1 + C_2|^2}{4S} \end{cases}, BC > 0; \quad \begin{cases} P_D^{G4U} = D + \dfrac{|C_1|^2}{D} \\ P_D^{S4R} = D + \dfrac{|C_1 + C_2|^2}{4D} \end{cases}, BC \leq 0. \tag{20}$$

From Eq. (20) we have

$$\begin{cases} P_S^{G4U} - P_S^{S4R} = \dfrac{|2C_1|^2 - |C_1 + C_2|^2}{4S}, BC > 0 \\ P_D^{G4U} - P_D^{S4R} = \dfrac{|2C_1|^2 - |C_1 + C_2|^2}{4D}, BC \leq 0 \end{cases}. \tag{21}$$

Then Eq. (19) will hold if $|2C_1|^2 - |C_1 + C_2|^2 \geq 0$. Obviously, this condition is not always tenable. Hence, despite better performance in some areas, G4U cannot improve S4R for every target area. To tackle this, the extended G4U (EG4U) will be developed in the following as an adaptive combination of G4U and DG4U.

3.6 EG4U: adaptive combination of G4U and DG4U

Combining $\mu = -1$ into Eqs. (12) and (15), DG4U surface and double-bounce scattering powers can be formulated as

$$\begin{cases} P_S^{DG4U} = S + \dfrac{|C_2|^2}{S}, BC > 0 \\ P_D^{DG4U} = D + \dfrac{|C_2|^2}{D}, BC \leq 0 \end{cases}. \tag{22}$$

Combining Eqs. (20) and (22), after some simple deduction, we obtain

$$\begin{cases} \dfrac{P_S^{G4U} + P_S^{DG4U}}{2} - P_S^{S4R} = \dfrac{|C_1 - C_2|^2}{4S} \geq 0, BC > 0 \\ \dfrac{P_D^{G4U} + P_D^{DG4U}}{2} - P_D^{S4R} = \dfrac{|C_1 - C_2|^2}{4D} \geq 0, BC \leq 0 \end{cases} \tag{23}$$

$$\begin{cases} P_S^{G4U} - P_S^{DG4U} = \dfrac{|C_1|^2 - |C_2|^2}{S}, BC > 0 \\ \\ P_D^{G4U} - P_D^{DG4U} = \dfrac{|C_1|^2 - |C_2|^2}{D}, BC \le 0 \end{cases}. \tag{24}$$

We can immediately obtain from Eq. (23) that

$$\begin{cases} \max\left\{P_S^{G4U}, P_S^{DG4U}\right\} \ge P_S^{S4R}, BC > 0 \\ \max\left\{P_D^{G4U}, P_D^{DG4U}\right\} \ge P_D^{S4R}, BC \le 0 \end{cases}. \tag{25}$$

From Eq. (24) we obtain

$$\begin{cases} \begin{cases} \max\left\{P_S^{G4U}, P_S^{DG4U}\right\} = P_S^{G4U}, BC > 0 \\ \max\left\{P_D^{G4U}, P_D^{DG4U}\right\} = P_D^{G4U}, BC \le 0 \end{cases}, BC_1 > 0 \\ \\ \begin{cases} \max\left\{P_S^{G4U}, P_S^{DG4U}\right\} = P_S^{DG4U}, BC > 0 \\ \max\left\{P_D^{G4U}, P_D^{DG4U}\right\} = P_D^{DG4U}, BC \le 0 \end{cases}, BC_1 \le 0 \end{cases} \tag{26}$$

where $BC_1 = |C_1| - |C_2|$. Eq. (26) just lays the foundation for EG4U:

$$\text{EG4U} := \mathcal{G}(\pm 1) = \begin{cases} \mathcal{G}(+1) = \text{G4U}, BC_1 > 0 \\ \mathcal{G}(-1) = \text{DG4U}, BC_1 \le 0 \end{cases}. \tag{27}$$

As the adaptive combination of G4U and DG4U, EG4U is also a special case of GG4U. So we denote it as $\mathcal{G}(\pm 1)$. By bringing $\mu = +1$ or $\mu = -1$ into Eqs. (12) and (15) based on the branch condition BC_1, we can achieve the scattering powers of four components in EG4U. Furthermore, from Eqs. (25) to (27), we have

$$\begin{cases} P_S^{EG4U} = \max\left\{P_S^{G4U}, P_S^{DG4U}\right\} \ge \left\{P_S^{S4R}, P_S^{G4U}, P_S^{DG4U}\right\}, BC > 0 \\ P_D^{EG4U} = \max\left\{P_D^{G4U}, P_D^{DG4U}\right\} \ge \left\{P_D^{S4R}, P_D^{G4U}, P_D^{DG4U}\right\}, BC \le 0 \end{cases}. \tag{28}$$

Compared with S4R and G4U, EG4U increases surface scattering in area where surface scattering is superior to double-bounce scattering and strengthens double-bounce scattering in area where double-bounce scattering is preferable to surface scattering. Therefore, EG4U achieves not only a nice improvement to S4R, but also an effective extension to G4U. This may make EG4U more suitable to the remote sensing of tsunami/earthquake. We will investigate this in Section 4. The procedure of EG4U is outlined in **Algorithm 1**.

Algorithm 1: EG4U

01: Input: $\langle[T]\rangle$
02: Conduct deorientation to $\langle[T]\rangle$ for $\langle[T']\rangle$
03: Compute helix power $P_C = 2|\text{Im}\{T'_{23}\}|H(T'_{33} - |\text{Im}\{T'_{23}\}|)$
04: Calculate branch condition BC
05: Determine volume scattering model based on branch condition
06: Obtain volume scattering power $P_V = (2T'_{33} - P_C)/2c$
07: Compute parameters S, D, C_1, and C_2, as well as branch condition BC_1
08: Implement *SPAN* reservation ruling based on $S + D$
09: if $S + D > 0$

10:   Adaptively select between G4U and DG4U based on BC_1
11:   if $BC_1 > 0$
12:      $C = C_1$
13:   else
14:      $C = C_2$
15:   end if
16:   Calculate surface scattering power P_S and double-bounce scattering power P_D according to BC
17:   if $BC > 0$
18:      $P_S = S + |C|^2/S, P_D = D - |C|^2/S$
19:   else
20:      $P_S = S - |C|^2/D, P_D = D + |C|^2/D$
21:   end if
22:   Implement nonnegative P_S and P_D ruling
23: else
24:   $P_S = P_D = 0, P_V = T'_{11} + T'_{22} + T'_{33} - P_C$
25: end if
26: Output: P_S, P_D, P_V, P_C

---

4. Monitoring of disaster by EG4U decomposition of ALOS-PALSAR images of 2011 Tohoku tsunami/earthquake

As indicated in Subsection 3.4, G4U and S4R represent two special forms of GG4U of equal status. Hence, G4U cannot fully improve S4R only if we ascend the status of G4U by combining the duality of G4U, i.e., DG4U and G4U together for EG4U. EG4U can adaptively strengthen the surface scattering and double-bounce scattering. Therefore, it may improve the competence and performance of G4U in the remote sensing of damages caused by earthquake/tsunami disaster. We demonstrate these in the following by decomposing the ALOS-PALSAR images of the 2011 great Tohoku tsunami/earthquake using EG4U.

4.1 Great Tohoku earthquake and tsunami

The great Tohoku earthquake is also known as the great Sendai earthquake or the great East Japan earthquake, which was a magnitude 9.0–9.1 (Mw) undersea megathrust earthquake off the coast of northeast Japan (the epicenter is shown in **Figure 2** as "⋆") that occurred on March 11, 2011, the most powerful earthquake ever recorded in Japan [34]. The earthquake triggered powerful tsunami, which swept the mainland of Japan, killed over 10,000 people (mainly through drowning), and damaged over 1,000,000 buildings (half of them are collapsed and even totally collapsed) [35].

4.2 Datasets

The Advanced Land Observing Satellite (ALOS) was launched in 2006 by the Japanese Space Agency (JAXA). It has three remote sensing payloads, i.e., the Panchromatic Remote-sensing Instrument for Stereo Mapping (PRISM) for digital elevation mapping, the Advanced Visible and Near Infrared Radiometer type 2 (AVNIR-2) for precise land coverage observation, and the Phased Array type L-band SAR (PALSAR) for all-day/all-weather land observation [36].

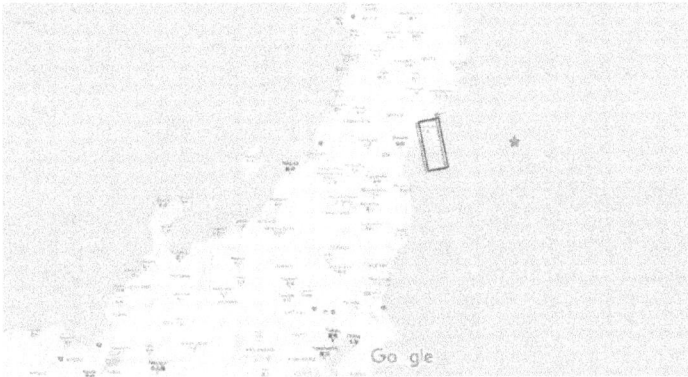

Figure 2.
Location of the great Tohoku tsunami/earthquake epicenter (★) and the ALOS-PALSAR footprint of the two selected fully polarimetric datasets (red rectangle, pre-event; blue rectangle, post-event).

Scene ID	Acquire data	Incidence angle[1]	Azimuth resolution	Ground-range resolution[2]
ALPSRP257090760	2010-11-21	23.802°	4.5 m	23.5 m
ALPSRP277220760	2011-04-08	23.836°	4.5 m	23.5 m

[1]*The incidence angle here indicates the incidence angle at the scene center.*
[2]*The ground-range resolution is defined as the slant-range resolution/sin(incidence angle) [9], while the slant-range resolution of the two datasets is both 9.5 m.*

Table 1.
ALOS-PALSAR datasets used in the experiment and their characteristics.

To demonstrate the capability of polarimetric remote sensing for damage monitoring, we choose two quad-polarization single-look complex-level 1.1 (ascending orbit) datasets acquired around Miyagi Prefecture, Japan, before and after the earthquake/tsunami with 138 days' temporal baseline, as summarized in **Table 1**. The ALOS-PALSAR footprint of the two datasets is shown in **Figure 2**.

4.3 Method

The flowchart of EG4U-based monitoring and evaluation of damages caused by tsunami/earthquake disaster is illustrated in **Figure 3**. We first co-register the two datasets based on the image features [37–40]. The boxcar filtering [9] is then carried out to both datasets to suppress the speckles. To ensure the pixel size in both image directions comparable, the window size for ensemble average is chosen as 2 pixels in ground-range direction and 12 pixels in azimuth direction, i.e., we integrate the scattering matrix $[S]$ of a total of 24 pixels for the estimation of a coherency matrix $\langle[T]\rangle$ in Eq. (2). From $\langle[T]\rangle$ we calculate the orientation angle θ according to Eq. (4) and implement the deorientation operation for the deoriented coherency matrix $\langle[T']\rangle$ according to Eq. (5). Finally, EG4U is used to decompose $\langle[T']\rangle$ to extract scattering powers P_S, P_D, P_V, and P_C and construct the RGB pseudo-color scattering power visualization result by encoding $\{R, G, B\}$ with $\{\sqrt{P_D}, \sqrt{P_V}, \sqrt{P_S}\}$. This process is executed on each cell of the two datasets until we obtain the complete pre- and post-event scattering power images shown in **Figure 4**, based on which we evaluate EG4U on monitoring of the tsunami/earthquake disaster in the following.

```
┌─────────────────────────────────────┐
│  Co-register the pre-event and       │
│  post-event datasets based on        │
│  image features                      │
└─────────────────────────────────────┘
```

| Ensemble average the pre-event scattering matrices $[S^{pre}]$ for the coherency matrix $\langle[T^{pre}]\rangle$ | Ensemble average the post-event scattering matrices $[S^{post}]$ for the coherency matrix $\langle[T^{post}]\rangle$ |

| Estimate orientation angle and execute deorientation operation to $\langle[T^{pre}]\rangle$ for $\langle[T^{pre'}]\rangle$ | Estimate orientation angle and execute deorientation operation to $\langle[T^{post}]\rangle$ for $\langle[T^{post'}]\rangle$ |

| Decompose $\langle[T^{pre'}]\rangle$ using EG4U for the scattering powers P_S^{pre}, P_D^{pre}, P_V^{pre}, and P_C^{pre} | Decompose $\langle[T^{post'}]\rangle$ using EG4U for the scattering powers P_S^{post}, P_D^{post}, P_V^{post}, and P_C^{post} |

| Construct the RGB color-coded pre-event scattering power image based on P_S^{pre}, P_D^{pre}, and P_V^{pre} | Construct the RGB color-coded post-event scattering power image based on P_S^{post}, P_D^{post}, and P_V^{post} |

| Compare and analyze the pre-event and post-event scattering power images for disaster monitoring |

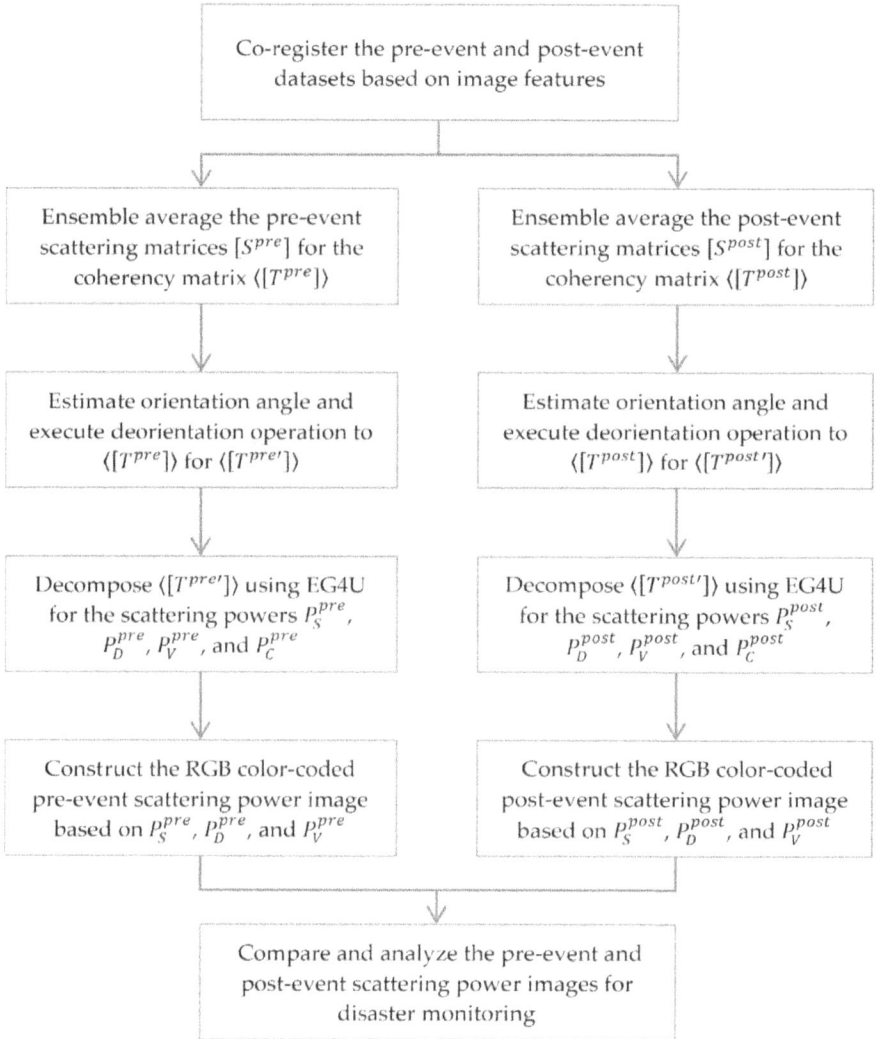

Figure 3.
Flowchart of EG4U-based monitoring of tsunami/earthquake disaster.

4.4 Evaluation and analysis

For better comparison and analysis, we also display the optical image of the study area obtained from ©Google Earth in **Figure 5**. Our intuitive impression of **Figure 4(a)** and **(b)** is their consistency and nice correspondence to the optical image. The blue color mainly appears in the water and land areas because of the dominant surface scattering there. The red color mainly arises in the urban area, such as the Ishinomaki City and Higashi-Matsushima City, with a large number of buildings. The ground and the vertical walls of buildings constitute the dihedral corner structures, which generally reflect the dominant double-bounce scattering. Mountain presents the green color, i.e., the dominant volume scattering. The well-developed branch and crown structures of trees on the mountain complicate the scattering process, depolarize the scattering wave, and show themselves as the complex mixed volume scattering in PolSAR image. Therefore, by color-coding the

Figure 4.
Color-coded scattering power image of the study area (a) before and (b) after the great Tohoku tsunami/ earthquake disaster. The framed patch regions A, B, and C are extracted for particular analysis.

scattering powers obtained by EG4U, we can achieve a nice discrimination of the ground objects.

Despite the consistency, we can also observe the obvious difference between the pre- and post-event scattering power images. A lot of red pixels in **Figure 4(a)** change to blue pixels in **Figure 4(b)**, particularly in the urban areas of Ishinomaki and Higashi-Matsushima, which illustrate the change from the dominant double-bounce scattering to the dominant surface scattering, denote the decrease of the dihedral structures, and indicate the collapse of buildings. Take Ishinomaki City framed in Patch A for instance; it is interesting to observe that the strong change mainly arises in the area by the seaside, while tiny change occurs in the area away from the coast. This finding is also validated by the corresponding optical images acquired before and after the event shown in **Figure 6(a)** and **(b)**. Therefore, the severe damages brought by the Tohoku tsunami/earthquake are probably mainly due to the flooding rather than the earthquake. Flooding from the Onagawa Bay and the Mangokuura Sea also swept the town of Onagawa framed in Patch B,

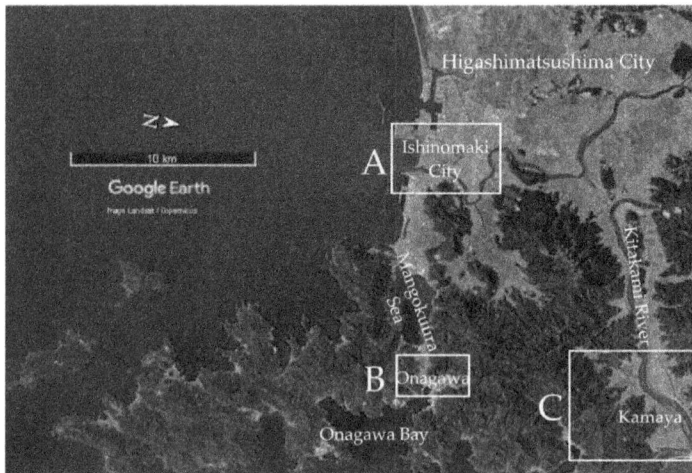

Figure 5.
Optical image of the study area obtained from ©Google earth. Particular attention is paid to the framed patch regions A, B, and C.

as shown in **Figure 6(c)** and **(d)** in terms of the pre- and post-event optical images. A large majority of red pixels of Patch B in **Figure 4(a)** change to blue pixels or even green pixels in **Figure 4(b)**, which indicates that nearly all the buildings in Onagawa were badly damaged by the flooding except for a few buildings constructed in high elevation. The collapsed buildings not only present the dominant surface scattering here, but also the dominant volume scattering because of the complex scattering in such mountain area. The biggest change caused by flooding appears in the area along the Kitakami River. Take the town of Kamaya framed in Patch C, for example, as shown in **Figure 6(e)**, besides several buildings, the most part of Kamaya is farmland. This area can be clearly distinguished from the Kitakami River in **Figure 4(a)** before the disaster. However, after the disaster, nearly all the land and buildings in Kamaya are flooded by the water from Kitakami River as shown in **Figure 6(f)**, which present in **Figure 4(b)** as the wide distribution of blue pixels and show the dominant surface scattering here. Therefore, by decomposing the pre- and post-event PolSAR datasets with EG4U to construct the color-coded scattering power images, we can achieve a simple but accurate monitoring of the damages caused by tsunami/earthquake disaster.

From the above analysis, we can obtain that flooding which resulted from tsunami is the main contributor to the severe damages in the 3.11 great Tohoku earthquake. The flooding destroyed the buildings and inundated the lands. All these damages present themselves in the polarization domain as the change of the dominant scattering mechanism from double-bounce scattering to surface scattering and in the image domain as the change of pixel color from red to blue. The boundary condition BC has been widely used in model-based decomposition as a crucial feature to discriminate surface scattering and double-bounce scattering [23, 24, 26–28]. As expressed in Eq. (14), $BC > 0$ indicates stronger surface scattering than double-bounce scattering, while $BC \leq 0$ denotes stronger double-bounce scattering than surface scattering. Therefore, besides the qualitative evaluation in terms of color, we can further achieve an quantitative evaluation of the damages by analyzing the dominant scattering according to BC. **Figure 7(a)** and **(b)** show the binary images of BC before and after the disaster, respectively. The white pixel denotes $BC > 0$, i.e., the dominant surface scattering, which mainly occupies the water and land areas, while the black one denotes $BC \leq 0$, i.e., the dominant double-bounce

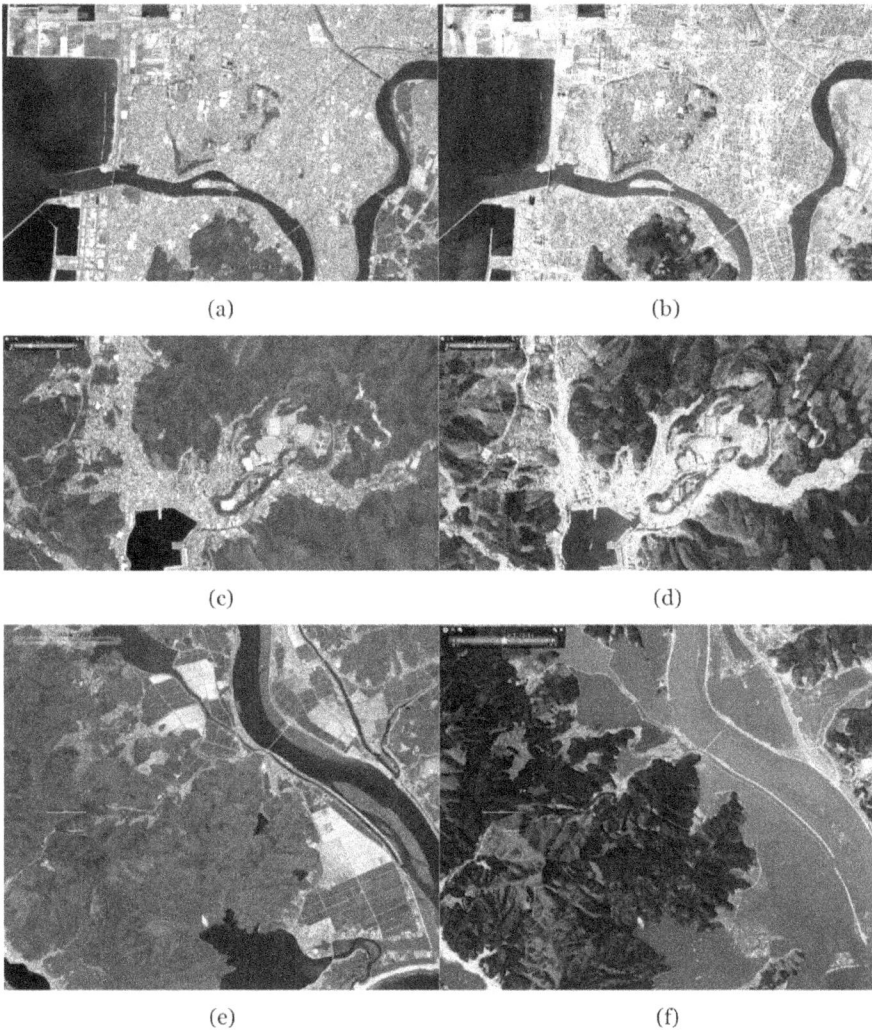

Figure 6.
Optical images of (first row, i.e. (a) and (b)) patch A, (second row, i.e. (c) and (d)) patch B, and (third row, i.e. (e) and (f)) patch C obtained from ©Google earth (first column, i.e. (a), (c), and (e)) before and (second column, i.e. (b), (d), and (f)) after the 3.11 great Tohoku tsunami/earthquake.

scattering, which mainly occupies the urban and mountain areas. Before disaster, the black pixels account for 15.1641% of the whole image, while this ratio decreases to 13.0785% after the disaster, i.e., the dominant scattering mechanism of about 2.0856% area of the scene is changed from double-bounce scattering to surface scattering. As shown in **Figure 7**, the change mainly arises in the urban area like the Ishinomaki City and Higashi-Matsushima City, in the land area like the town of Kamaya, as well as in the water area like the Mangokuura Sea, Onagawa Bay, and Kitakami River. This further provides us a consistently quantitative evaluation of the damages. All these demonstrate the importance and value of polarimetric microwave remote sensing technique in the monitoring of tsunami/earthquake damages.

Singh et al. [28] indicated that G4U could enhance double-bounce scattering over urban area while strengthen surface scattering contribution over water and land area. This establishes G4U the state-of-the-art four-component scattering

(a)   (b)

Figure 7.
Binary display of the branch condition BC extracted from (a) pre- and (b) post-event ALOS-PALSAR datasets. The white pixels correspond to BC > 0, while the black pixels denote BC ≤ 0.

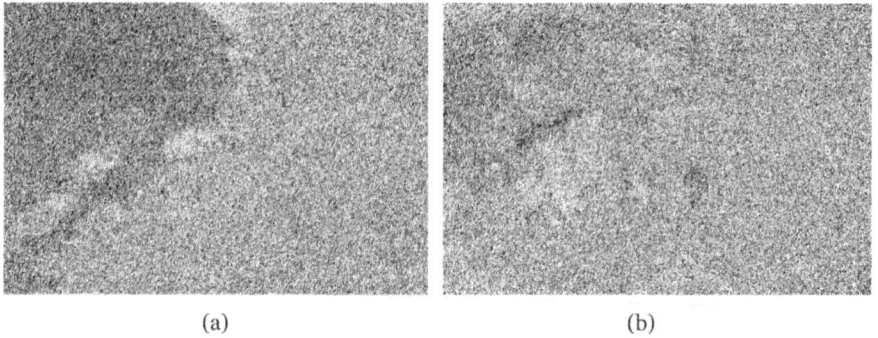

(a)   (b)

Figure 8.
Binary display of the branch condition BC_1 extracted from (a) pre- and (b) post-event ALOS-PALSAR datasets. The white pixels correspond to $BC_1 > 0$, while the black pixels denote $BC_1 \leq 0$.

power decomposition and enables its wide application to the remote sensing of forestry, agriculture, wetland, snow, glaciated terrain, earth surface, manmade target, environment, and damages caused by earthquake, tsunami, and landslide [29, 30]. Nevertheless, the rigorous derivation in Eq. (21) validates that G4U cannot always enhance the double-bounce scattering nor strengthen the surface scattering power unless we adaptively integrate G4U and its duality, i.e., DG4U, for EG4U based on another boundary condition BC_1. As expressed in Eq. (27), G4U is selected only when $BC_1 > 0$; otherwise, we should turn to DG4U. The binary images **Figure 8(a)** and **(b)** further show the pre- and post-event BC_1, respectively, where the white pixels (i.e., $BC_1 > 0$) indicate the area where G4U operates and the black pixels (i.e., $BC_1 \leq 0$) give the area where DG4U operates. The white pixels account for 46.4260% of the pre-event image, which conveys that G4U achieves better result than S4R only for 46.4260% area. As for the rest 53.5740% area, we should resort to DG4U for improvement. The ratio of white pixels increases to 49.5247% after the disaster. Nevertheless, there are still half a little more areas where G4U will underestimate the surface or double-bounce scattering. If we adopt G4U in this area to evaluate damages caused by tsunami/earthquake, the reduced double-bounce scattering from G4U may lead to the underestimation of building scale and overestimation of damage level. EG4U can adaptively increase the surface scattering or double-bounce scattering. Hence, it definitely improves the competence and performance of G4U in the remote sensing of damages caused by earthquake/tsunami.

5. Conclusion

Flooding is the main contributor to the severe damages in the great Tohoku tsunami/earthquake. It destroyed the buildings and inundated the lands by the seaside. All these damages present themselves in the polarization domain as the change of the dominant scattering mechanism from double-bounce scattering to surface scattering and in the image domain as the change of pixel color from red to blue. The color-coded scattering power image is very useful and powerful in the qualitative evaluation of damages. The boundary condition BC further enables a nice quantitative evaluation of disaster. The unitary transformation in G4U adds a T_{13}-related but redundant balance equation to the original self-contained equation system. The general solution enables a generalized G4U, while G4U just represents a special form. The strict derivation conveys that G4U cannot always strengthen the double-bounce scattering in urban area nor strengthen the surface scattering in water or land area unless we adaptively combine G4U and its duality for EG4U. Experiment on the ALOS-PALSAR datasets of 2011 great Tohoku tsunami/earthquake demonstrates not only the outperformance of EG4U but also the effectiveness of polarimetric remote sensing in the qualitative monitoring and quantitative evaluation of tsunami/earthquake damages. Efficient and accurate monitoring and assessment are of crucial importance for the fast response, management, and mitigation of the disasters. The all-day and all-weather working capacity is a significant advantage of microwave remote sensing. Polarimetric remote sensing is an effective technique in the discrimination and recognition of ground objects.

Acknowledgements

This work was supported in part by the National Natural Science Foundation of China under Grant No. 41871274 and No. 61971402 and by the Strategic High-Tech Innovation Fund of Chinese Academy of Sciences under Grant CXJJ19B10.

Conflict of interest

The authors declare no conflict of interest.

Notes

Sections 2 and 3 of this chapter are extracted from a journal paper of the authors submitted to IEEE Transactions on Geoscience and Remote Sensing on June 07, 2017. The paper is still under review at the time of publication of this chapter. For more details about the paper, please refer to Reference [30].

Author details

Dong Li[1,2*], Yunhua Zhang[1,2*], Liting Liang[1,2], Jiefang Yang[1] and Xun Wang[1,2]

1 Key Laboratory of Microwave Remote Sensing, National Space Science Center, Chinese Academy of Sciences, Beijing, China

2 University of Chinese Academy of Sciences, Beijing, China

*Address all correspondence to: lidong@mirslab.cn and zhangyunhua@mirslab.cn

IntechOpen

References

[1] Mokhtari M, editor. Tsunami-A Growing Disaster. London: IntechOpen; 2011. p. 232. DOI: 10.5772/922

[2] Morner N-A, editor. The Tsunami Threat–Research and Technology. London: IntechOpen; 2011. p. 714. DOI: 10.5772/573

[3] Mokhtari M, editor. Tsunami. London: IntechOpen; 2016. p. 164. DOI: 10.5772/61999

[4] Marghany M, editor. Advanced Remote Sensing Technology for Synthetic Aperture Radar Applications, Tsunami Disasters, and Infrastructure. London: IntechOpen; 2019. p. 167. DOI: 10.5772/intechopen.78525

[5] Li D, Zhang Y. Adaptive model-based classification of PolSAR data. IEEE Transactions on Geoscience and Remote Sensing. 2018;**56**(12):6940-6955. DOI: 10.1109/TGRS.2018.2845944

[6] Li D, Zhang Y. Random similarity-based entropy/alpha classification of PolSAR data. IEEE Journal of Selected Topics in Applied Earth Observations and Remote Sensing. 2017;**10**(12): 5712-5723. DOI: 10.1109/JSTARS.2017. 2748234

[7] Li D, Zhang Y. Unified Huynen phenomenological decomposition of radar targets and its classification applications. IEEE Transactions on Geoscience and Remote Sensing. 2016; **54**(2):723-743. DOI: 10.1109/TGRS. 2015.2464113

[8] Li D, Zhang Y. Random similarity between two mixed scatterers. IEEE Geoscience and Remote Sensing Letters. 2015;**12**(12):2468-2472. DOI: 10.1109/ LGRS.2015.2484383

[9] Lee J-S, Pottier E. Polarimetric Radar Imaging: From Basics to Applications. Boca Raton: CRC Press;

2009. p. 422. DOI: 10.1201/97814200 54989

[10] Cloude SR. Polarisation Applications in Remote Sensing. Oxford: Oxford University Press; 2010. p. 453. DOI: 10.1063/1.3502550

[11] van Zyl JJ, Kim Y. Synthetic Aperture Radar Polarimetry. Hoboken: John Wiley & Sons, Inc.; 2011. p. 312. DOI: 10.1002/9781118116104

[12] Yamaguchi Y. Disaster monitoring by fully polarimetric SAR data acquired with ALOS-PALSAR. Proceedings of the IEEE. 2012;**100**(10):2851-2860. DOI: 10.1109/JPROC.2012.2195469

[13] Sato M, Chen S-W, Satake M. Polarimetric SAR analysis of tsunami damage following the march 11, 2011 East Japan earthquake. Proceedings of the IEEE. 2012;**100**(10):2861-2875. DOI: 10.1109/JPROC.2012.2200649

[14] Watanabe M, Motohka T, Miyagi Y, Yonezawa C, Shimada M. Analysis of urban areas affected by the 2011 off the Pacific coast of Tohoku earthquake and tsunami with L-band SAR full-polarimetric mode. IEEE Geoscience and Remote Sensing Letters. 2012;**9**(3): 472-476. DOI: 10.1109/LGRS.2011. 2182030

[15] Li X, Zhang L, Guo H, Sun Z, Liang L. New approaches to urban area change detection using multitemporal RADARSAT-2 polarimetric synthetic aperture radar (SAR) data. Canadian Journal of Remote Sensing. 2012;**38**(3): 253-255. DOI: 10.5589/m12-018

[16] Singh G, Yamaguchi Y, Boerner W-M, Park S-E. Monitoring of the March 11, 2011, off-Tohoku 9.0 earthquake with super-tsunami disaster by implementing fully polarimetric high-resolution POLSAR techniques.

Proceedings of the IEEE. 2013;**101**(3): 831-846. DOI: 10.1109/JPROC.2012. 2230311

[17] Chen S-W, Sato M. Tsunami damage investigation of built-up areas using multitemporal spaceborne full polarimetric SAR images. IEEE Transactions on Geoscience and Remote Sensing. 2013;**51**(4):1985-1997. DOI: 10.1109/TGRS.2012.2210050

[18] Li N, Wang R, Deng Y, Liu Y, Wang C, Balz T, et al. Polarimetric response of landslides at X-band following the Wenchuan earthquake. IEEE Geoscience and Remote Sensing Letters. 2014;**11**(10):1722-1726. DOI: 10.1109/LGRS.2014.2306820

[19] Chen S-W, Wang X-S, Sato M. Urban damage level mapping based on scattering mechanism investigation using fully polarimetric SAR data for the 3.11 East Japan earthquake. IEEE Transactions on Geoscience and Remote Sensing. 2016;**54**(12):6919-6929. DOI: 10.1109/TGRS.2016.2588325

[20] Ji Y, Sumantyo JTS, Chua MY, Waqar MM. Earthquake/tsunami damage level mapping of urban areas using full polarimetric SAR data. IEEE Journal of Selected Topics in Applied Earth Observations and Remote Sensing. 2018;**11**(7):2296-2309. DOI: 10.1109/JSTARS.2018.2822825

[21] Zhai W, Huang C, Pei W. Two new polarimetric feature parameters for the recognition of the different kinds of buildings in earthquake-stricken areas based on entropy and eigenvalues of PolSAR decomposition. Remote Sensing. 2018;**10**(10):1613. DOI: 10.3390/rs10101613

[22] Cloude SR, Pottier E. A review of target decomposition theorems in radar polarimetry. IEEE Transactions on Geoscience and Remote Sensing. 1996;**34**(2):498-518. DOI: 10.1109/36.485127

[23] Freeman A, Durden SL. A three-component scattering model for polarimetric SAR data. IEEE Transactions on Geoscience and Remote Sensing. 1998;**36**(3):963-973. DOI: 10.1109/36.673687

[24] Yamaguchi Y, Moriyama T, Ishido M, Yamada H. Four-component scattering model for polarimetric SAR image decomposition. IEEE Transactions on Geoscience and Remote Sensing. 2005;**43**(8):1699-1706. DOI: 10.1109/TGRS.2005.852084

[25] Zhu F, Zhang Y, Li D. A novel deorientation method in PolSAR data processing. Remote Sensing Letters. 2016;**7**(11):1083-1092. DOI: 10.1080/2150704X.2016.1217438

[26] Yamaguchi Y, Sato A, Boerner W-M, Sato R, Yamada H. Four-component scattering power decomposition with rotation of coherency matrix. IEEE Transactions on Geoscience and Remote Sensing. 2011;**49**(6):2251-2258. DOI: 10.1109/TGRS.2010.2099124

[27] Sato A, Yamaguchi Y, Singh G, Park S-E. Four-component scattering power decomposition with extended volume scattering model. IEEE Geoscience and Remote Sensing Letters. 2012;**9**(2):166-170. DOI: 10.1109/LGRS.2011.2162935

[28] Singh G, Yamaguchi Y, Park S-E. General four-component scattering power decomposition with unitary transformation of coherency matrix. IEEE Transactions on Geoscience and Remote Sensing. 2013;**51**(5):3014-3022. DOI: 10.1109/TGRS.2012.2212446

[29] Li D, Zhang Y, Liang L. A Concise Survey of G4U. 2019. Available from: https://arxiv.org/ftp/arxiv/papers/1910/1910.14323.pdf [Accessed: 28 December 2019]

[30] Li D, Zhang Y, Liang L. A mathematical extension to the general

four-component scattering power decomposition with unitary transformation of coherency matrix. IEEE Transactions on Geoscience and Remote Sensing. 2020:1-18. Under review

[31] Sinclair G. The transmission and reception of elliptically polarized waves. Proceedings of the IRE. 1950;**32**(2): 148-151. DOI: 10.1109/JRPROC.1950. 230106

[32] Huynen JR. Phenomenological theory of radar targets [thesis]. Delft: Delft University of Technology; 1970

[33] Cloude SR, Pottier E. An entropy based classification scheme for land applications of polarimetric SAR. IEEE Transactions on Geoscience and Remote Sensing. 1997;**35**(1):68-78. DOI: 10.1109/36.551935

[34] Wikipedia. 2011 Tohoku Earthquake and Tsunami. 2019. Available from: https://en.wikipedia. org/wiki/2011_T%C5%8Dhoku_ earth quake_and_tsunami. [Accessed: 28 December 2019]

[35] National Police Agency of Japan. Police Countermeasures and Damage Situation associated with 2011 Tohoku district-off the Pacific Ocean Earthquake. 2019. Available from: https://www.npa.go.jp/news/other/ earthquake2011/pdf/higaijokyo_e.pdf [Accessed: 28 December 2019]

[36] JAXA Earth Observation Research Center. About ALOS–Overview and Objectives. 2019. Available from: https://www.eorc.jaxa.jp/ALOS/en/ about/about_index.htm [Accessed: 28 December 2019]

[37] Li D, Zhang Y. A novel approach for the registration of weak affine images. Pattern Recognition Letters. 2012; **33**(12):1647-1655. DOI: 10.1016/j. patrec.2012.04.009

[38] Li D, Zhang Y. A fast offset estimation approach for InSAR image subpixel registration. IEEE Geoscience and Remote Sensing Letters. 2012;**9**(2): 267-271. DOI: 10.1109/LGRS.2011. 2166752

[39] Li D, Zhang Y. On the appropriate feature for general SAR image registration. Proceedings of SPIE. 2012; **8536**:8536X. DOI: 10.1117/12.970520

[40] Li D, Zhang Y. The appropriate parameter retrieval algorithm for feature-based SAR image registration. Proceedings of SPIE. 2012;**8536**:8536Y. DOI: 10.1117/12.970522

Chapter 3

Dealing with Local Tsunami on Pakistan Coast

Ghazala Naeem

Abstract

Tsunami originating from a local source can arrive at Pakistan coastline within minutes. In the absence of a comprehensive and well-coordinated management plan, the fast-approaching tsunami might wreak havoc on the coast. To combat such a threat, a wide range of short- and long-term mitigation measures are needed to be taken by several government and private sector organizations as well as security agencies. Around 1000-km coastline is divided administratively into two provinces of Baluchistan and Sindh and further into seven districts. Most of the coastal communities were severely affected by an earthquake of magnitude 8+ on 28 November 1945 followed by a devastating tsunami. In contrast to the level of posed hazard and multiple-fold increase in vulnerabilities since then, the risk mitigation efforts are trivial and least coordinated. It is important to provide stakeholders with a set of prerequisite information and guidelines on standardized format to develop their organizational strategies and course of action for earthquake and tsunami risk mitigation in a well-coordinated manner, from local to the national level.

Keywords: local tsunami, hazard and risk assessment, mitigation, preparedness, standardized format, stakeholders' coordination

1. Introduction

Tsunami being a less frequent hazard has not yet gained due attention in the national hazard mitigation and preparedness program within Pakistan. However, disastrous impacts of 1945 Makran Tsunami, which occurred in the Arabian Sea merely 70 years ago, cannot be ignored and urge need of comprehensive and sustained tsunami resilience efforts.

In recent decades, 2004 Indian Ocean and 2011 Japan Tsunamis have revealed destructing powers of tsunami and the level of unpreparedness with regard to hazard assessment, warning and response planning, public awareness, mitigation, and research, not only of developing but developed countries as well.

Since 2006, in the aftermath of 2004 Indian Ocean Tsunami, significant efforts have been made in the country; however, there is much more to do for developing tsunami-resilient communities in Pakistan.

There are several multi-tiered stakeholders having inter-reliant responsibilities and mandates for earthquake and tsunami risk reduction, working in the coastal region of Pakistan. There is a need to support those stakeholders in dealing with tsunami and earthquake risks in a well-coordinated and comprehensive manner. This chapter recommends policy guidelines for determining strategic significance of posed tsunami threat to the Pakistan coast and ascertaining the underlying risks in comparison to the current capacities and preparedness measures. The chapter also suggests desirable research work, establishing timely warning system and structural and nonstructural mitigation measures including effective outreach to the public level and international coordination to combat local tsunami threat.

2. Recommended policy guidelines for local tsunami

The Arabian Sea region is threatened by earthquake and tsunami hazards, mainly because of the presence of the Makran subduction zone (MSZ). An earthquake of magnitude 8+ had wreaked havoc along the Pakistan coastline on 28 November 1945 followed by a devastating tsunami. In contrast to the level of posed threat and multiple-fold increase in vulnerabilities since then, the risk mitigation efforts are trivial and least coordinated. There is need for stakeholders to provide a set of prerequisite information to develop their organizational strategies and course of action for earthquake and tsunami risk mitigation in a well-coordinated manner, from local to the national level. Most important and immediate tasks include:

- Develop standardized and coordinated tsunami hazard and risk assessments for all coastal regions of Sindh and Baluchistan provinces.

- Improve tsunami and seismic sensor data, infrastructure, and standard operating procedures (SOPs) for better tsunami detection and warning.

- Enhance tsunami forecast and warning dissemination capability along the coastline.

- Promote the development of model mitigation measures, and encourage communities to adopt resilient construction, critical facilities protection, and land-use planning practices to reduce the impact of future tsunamis.

- Increase outreach to all communities, including all demographics of the at-risk population, to raise awareness, improve preparedness, and encourage the development of tsunami response plans.

- Develop a strategic plan for earthquake- and tsunami-related research especially within Arabian Sea region.

The required mitigation measures in a standardized manner are divided into three main categories including hazard assessment, risk evaluation, and mitigating measures to guide national level stakeholders in developing a long-term comprehensive tsunami response and risk reduction plan for Pakistan.

Establishing a technical committee to perform a role of central coordination and advisory under the National Disaster Management Authority can be supportive for stakeholders interested in and mandated for planning and implementation of earthquake and tsunami response and preparedness measures.

3. Assessing tsunami hazard

This section is based on the review of scientific and historical evidences of various potential sources of tsunamis which have affected and are likely to affect the Arabian Sea region and are described below.

3.1 Potential tsunami sources

3.1.1 Subduction zone earthquake

The most important source of earthquake-generated fast-approaching tsunamis (a local tsunami) in the Arabian Sea is the Makran subduction zone (**Figure 1**) adjacent to the coasts of Iran and Pakistan [1]. Recent event, known as 1945 Makran Tsunami, was caused by the earthquake in eastern part of this zone.

Another potential subduction zone lies from the northern tip of the Bay of Bengal, through the western margin of the Andaman Sea, and skirting the southern coasts of Sumatra, Java, and the islands of Lesser Sunda and is underlain geologically as Sunda subduction zone [1]. However, a tsunami generated by a potential earthquake event within Sunda subduction zone may reach Pakistan coast in hours, categorized as distant tsunami. A devastating event like 2004 Indian Ocean tsunami created disturbance on Pakistan coast after several hours of the incident occurred near Indonesia.

It is difficult to evaluate the accurate level of tsunami hazard these subduction zones pose for near and distant regions. The record and likelihood of earthquake occurrence in these zones and the implications for tsunami generation are the only basis for such estimations.

3.1.2 Submarine landslides

Within the Arabian Sea region, submarine landslides have the potential to produce large, local tsunamis owing to steep seafloor slopes and rapid sedimentation.

Tsunami waves ($\leq$1 m) were observed in the Arabian Sea on 24 September 2013 along several beaches in Oman and Pakistan during low tide period. The event was caused by a secondary effect of an earthquake of magnitude 7.7, which occurred

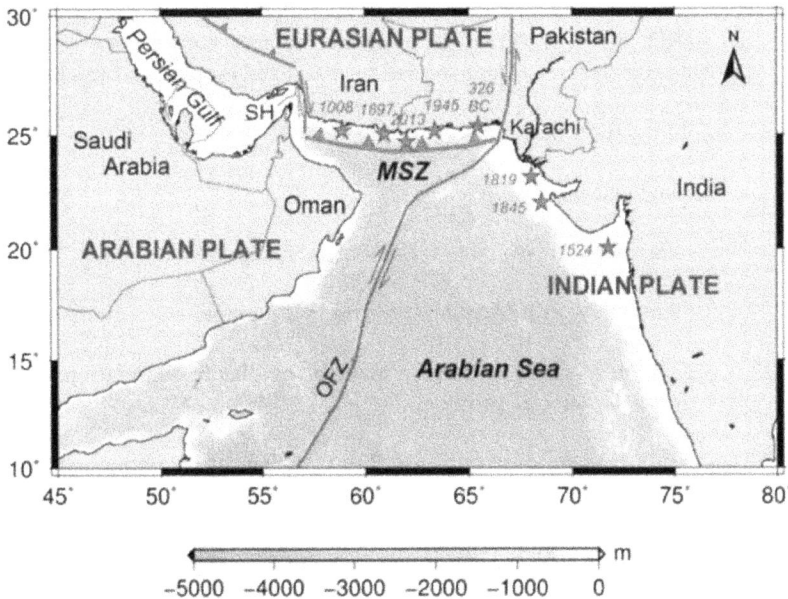

Figure 1.
General location map of the Makran subduction zone (MSZ) at the northwestern Indian Ocean showing locations of past tsunamis in the region. Source tsunami risk, preparedness and warning system in Pakistan by Heidarzadeh [2].

inland in southwestern Pakistan at 11.29.47 UTC (local time is UTC +5) on the same day, but after several hours, the earthquake's epicenter was a couple of hundred kilometers inland. Hoffman et al. [3] suggest the waves must have been triggered by a submarine landslide.

3.2 Guidelines for tsunami hazard assessment

Tsunami hazard assessment at local level is important to understand the locally imposed threat and to deal with the posed hazard accordingly. The creation of local hazard maps is a key step in the tsunami risk assessment procedure and is needed to:

- Develop evacuation plans

- Land-use planning within a defined coastal management area

- Determine the exposure parameters that will be used in the assessment of vulnerability of the coastal community and of their supporting assets and systems

The organizations, institutions, and experts mandated or interested to conduct directly or support to carry out earthquake and/or tsunami hazard assessment along the Pakistan coastline are suggested to follow standard parameters or guidelines, led by any national agency like the National Disaster Management Authority (NDMA) such as:

i. Local tsunami hazard maps are usually developed from specified tsunami event scenarios. The parameters defined in Multi-Hazard Vulnerability and Risk Assessment guidelines of the NDMA available at http://ndma.gov.pk/publications/MHVRA%202017.pdf are used [4]. Other technical details for modeling tsunami hazard, for example, input data sources, modeling tools, and criteria used by different research institutions and experts, can also be reviewed by the NDMA to make integrated tsunami hazard assessment of the Pakistan coastline on a standardized format. Final output products include:

   a. Inundation maps

   b. Flow velocity profiles

   c. Warning time available for emergency response

   d. Debris flow profiles at least for ports if possible

   e. Comprehensive hazard map made by overlaying inundation maps and flow velocity profiles

ii. The NDMA take the core responsibility of coordination among agencies and implementing partners to develop and facilitate regular update of hazard database. The NDMA collaboration is also required for statistics and information acquisition and sharing among stakeholders, developing and publishing such maps on standardized and easily understandable (for general public) format.

iii. Interagency coordination for sharing data and information required for comprehensive hazard assessment should be conducted by the NMDA and PDMAs, for example, onshore, offshore surface data, geological and meteorological scientific information, census and building records, and satellite images and archival records. The NDMA being the central focal organization for disaster management can play a vital role to facilitate data and information sharing among organizations and also with researchers.

iv. Tsunami caused by undersea landslides should also be accounted for more reliable hazard assessment process.

v. All possible impacts of any future tsunami event should also be studied and modeled in details. For example, huge quantities of debris brought onshore itself can be a major hazard for ports, fishing harbors, and local environment at any specific location.

4. Assessing tsunami risk

Using the outputs from the hazard assessment, disaster, emergency managers, and other relevant organizations (NDMA, PDMAs, and DDMAs of Baluchistan and Sindh Provinces) will need to create a community asset database of maps showing the distribution of population, buildings, infrastructure, and environmental assets in relation to the information on various hazard exposure parameters (inundation limit, run-up, depth of water, proximity to open coast, inundation and drainage flow velocities, etc.) for a particular earthquake and tsunami hazard scenario [1].

4.1 Guidelines for tsunami risk assessment

i. In addition to following the NDMA's MHVRA guidelines, tsunami vulnerability and capacity assessments are to be carried out in detail. Such risk maps should be interpreted in standardized format of "well-defined categories of risk" from the national to local level.

ii. Hazard maps developed under the proposed guidelines of the previous section can be used to incorporate vulnerability maps and finally produce tsunami risk maps, and those should be communicated to all stakeholders in a systematic way.

iii. Levels of risk are presented in geospatial ways: maps showing the extents of areas with defined "risk categories" as high, medium, and low levels of estimated risks as per the NDMA's MHVRA guidelines.

iv. Coastal cities being hub of economic activities having national life line infrastructure like ports are also densely populated. The NDMA and other relevant national and provincial agencies should encourage research institutions and experts to prioritize conduction of vulnerability and risk assessments of urban areas on priority basis and its incorporation in the national database. For example, detailed exposure database of 1- to 3-km-wide coastal belt in urban, semi-urban, and rural settlements can be maintained and updated annually by the provincial and district disaster management authorities.

v. To assure uniformity and speed in the risk assessment process, the following data of interest should be maintained for areas in close proximity to the sea (as suggested above, i.e., 1- to 3-km-wide belt along the coast can be surveyed on priority). Detailed survey should be carried out along the coastal areas to collect data such as:

- Census data (population distribution, income, and other statistics such as age, occupation, disability, education)

- Building classification, construction materials and techniques, ground level elevation

- Critical infrastructure (roads, water, power, sewerage, emergency facilities)

- Economic zones and location (business sectors, industry, ports)

- Environmental services/inventory

The abovementioned information include all required level sof onshore, offshore surface data, geological and meteorological scientific information, census and building records, satellite images, archival records, and organizational capacity (to facilitate and contribute in earthquake and tsunami emergency response and preparedness).

vi. Data and information collection, for reliable and authenticated coastal earthquake and tsunami risks assessment, can be acquired by all the agencies and organization mandated to collect and maintain such database, whereas the NDMA can play a vital role of coordination and support for essential data sharing among agencies and with experts by developing data sharing protocol.

5. Managing tsunami risk

This section covers guidelines for effective earthquake and tsunami risk reduction measure to strengthen coastal communities and infrastructure aiming to reduce impacts of any devastating event in the future.

5.1 Early warning system

Is it critically important to assess whether the current early warning system and practices are effective for the posed tsunami threat to Pakistan coastal communities [5]? A review of critical issues that hindered the efficient and timely operation of early warning systems has led to the identification of four elements [1]:

- Implementation of technically oriented early warning systems, without taking into consideration or without conducting risk assessment

- Weaknesses in monitoring and forecasting of potentially catastrophic events

- Weaknesses in the emission of warnings or in ensuring that warnings reach vulnerable communities

- Weaknesses in local capacities to respond to a warning and to a potentially catastrophic event

5.1.1 Current status

The National Seismic Monitoring Tsunami Early Warning Center (NSMTEWC) of the Pakistan Meteorological Department (PMD) is capable of issuing warning bulletins and messages to identified stakeholders including disaster management authorities, concerned provincial and district governments, and media within 13 min as specified in laid down standard operating procedures [5]. However, there is a lack of further downstream time bound SOPs (13 min onward with reference to PMD's SOPs) of other stakeholders (e.g., disaster management authorities, emergency services, provincial and local governments) to ensure the warning information and messages are communicated to all vulnerable coastal communities and, if needed, to adopt evacuation procedures that are timely completed within available lead time.

5.1.2 Guidelines for effective early warning system

The NDMA being the central coordinating agency of disaster management in Pakistan can take a lead and engage relevant organization including PDMA Sindh, PDMA Balochistan, Army, Pakistan Navy, Pakistan Coast Guards, Marine Security Agency, port authorities, and police to develop consensus on technical issues, set required protocols, and monitor progress on the policy guidelines mentioned below:

    i. End-to-end time bound synergized SOPs for the dissemination of tsunami warning to be developed involving all stakeholders living in remote coastal villages including islands and creeks.

    ii. Protocol among all national organizations capable of communication (using one or more communication networks such as satellite, HF/VHF, radio, or any wired or wireless network) should be established and made accessible to coastal communities (on- and offshore).

    iii. Develop and enforce SOPs and procedures to ensure that all tsunami and earthquake detection, forecasting, warning communication, and dissemination network/equipment must be kept in operational condition by the organization in charge of the asset. Such equipment and network should be installed and maintained with earthquake-resistant features and techniques.

    iv. Stakeholders and organizations responsible for burden sharing of early warning dissemination can be involved in practicing procedures and equipment operation tests, collectively, at least once a year or on an agreed schedule as a full-scale tsunami exercise in coordination with the NDMA.

    v. Individual government organizations involved in early warning chain and emergency response to arrange tabletop and functional test exercises at least twice a year on a feasible schedule, in coordination with NDMA and/or concerned PDMAs.

    vi. Official early warning bulletins should be adapted as easily understandable public messages by relevant PDMA in coordination with the PMD Tsunami Center. The NDMA shall provide central coordination to maintain uniformity and standardization of the public messages.

vii. Early warning bulletins and messages shall be tested for their level of understanding through a manageable size of survey after each simulation, exercise or drill involving the general public or at least schools in target areas.

viii. Strengthening of information sharing mechanism among Regional Tsunami Watch Providers and National Tsunami Warning Centers to better receive information and advice to complement national data stream including seismic, sea level, and other geophysical data networks.

5.2 Evacuation planning

Subject to the assessed level of risk in respect of a tsunami event, disaster management authorities and emergency responders should prioritize establishing and implementing a strategic plan (considering available lead time and resources at local level) for the effective and orderly evacuation of the exposed population.

Evacuation planning in each coastal area is directly related to the:

- Geographical size of the management area

- Assessed hazards and vulnerabilities

- Topography

- Demographics

- Size and density of the population

- Number of agencies involved in the planning process

- Resources available

Vulnerability maps derived from the inundation maps (in Section 4) provide key information for evacuation planning. Either voluntary or mandatory evacuation, both can place a significant burden on the resources and emergency managers in terms of caring for the displaced people (**Figures** 2–5).

Figure 2.
Pedestrian tsunami evacuation route at Gwadar City, Baluchistan Province. (Left) Red arrows show starting and ending point of the evacuation route. (Right) Tsunami evacuation route, more than 600 steps of a stairs designed with landing at several points to facilitate pedestrian evacuation leading to the proposed evacuation site at the top of the Koh-e-Batil (a 450 high mountain) by the district disaster management authority Gwadar. Photo by Ghazala Naeem.

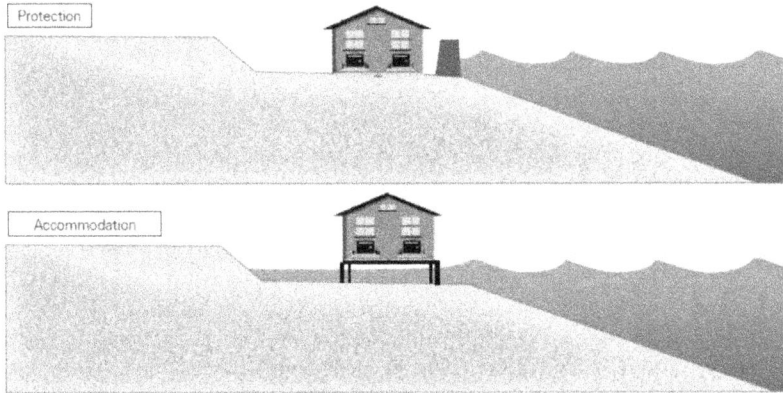

Figure 3.
Application of engineering solutions of building a protection wall and accommodating the building by raise on stilts to safeguard against maximum tsunami wave and inundation height. Source: IOC Manual 52 [1].

Figure 4.
One of the strategic options for tsunami mitigation is "retreat" with reference to general maximum tsunami wave and inundation height. Source: IOC Manuals and Guides 52 [1].

Figure 5.
Remote coastal communities settled on small islands in the Indus Creek Sindh Province. Photo by Ghazala Naeem.

5.2.1 Guidelines for tsunami evacuation planning

Disaster management authority of each coastal district, in coordination with concerned stakeholders, should prepare evacuation plans with the following aspects to be addressed:

    i. Identify "at-risk" people/communities who may require evacuation (either through risk assessment (Section 4)). It is recommended that authorities proceed with mapping based on current locally available information and indigenous knowledge and not wait for the perceived required scientific knowledge. Zone boundary definition can then be refined as knowledge improves, over time.

ii. Safe evacuation sites or buildings should be identified, clearly marked, and communicated to locals based on the perceived hazard analysis, for example, possible earthquake shaking, inundation height and extent, etc. Such sites and buildings should be pre-examined for safety, security, required space, and facilities to cater for the expected number of evacuees.

iii. Maps depicting tsunami evacuation zones, escape routes, and tsunami safe areas should be available for display at workplaces, public gathering areas and buildings, holiday homes, and tourist facilities. Particularly, display in all areas subjected to tsunami risk.

iv. Well-placed evacuation signage (in nationally agreed standardized format) with local perspective is critically important, for example, safety instructions and signage (natural signs) for tsunami events, identification of dangerous areas, safe sites, routes to reach evacuation sites, and other important messages.

v. Define conditions under which an evacuation may be necessary.

vi. Elaborate command, control, and coordination instructions (including designation of officials who are authorized to order an evacuation).

vii. Warning instructions should be issued to the media, public, and businesses.

viii. Procedures for assisting special categories of evacuees (e.g., vulnerable communities with least communication networks, elderly, children, physically challenged people, school students, patients at hospitals, etc.).

ix. Specific plans and procedures that address:

- The circumstances of the emergency

- Transportation (e.g., arrangements in areas where pedestrian evacuation is not possible or for the patients, etc.)

- Dealing with community that disregards mandatory evacuation

- The evacuation of specific locations (UC and village level) facilities (ports, hospitals, schools, large industrial setup, atomic reactors, security agencies set-up) and evacuation routes

x. Means of accounting for evacuees (and registration).

xi. Welfare support for evacuees; designated reception areas for vulnerable groups like unattended children, elderly, patients, physically challenged people, etc.

xii. Security of evacuated areas.

xiii. Procedures for the return of evacuees.

xiv. A consistent plan to facilitate common public understanding across communities about tsunami evacuation zones, maps, tsunami evacuation signage, and tsunami response actions.

xv. Maintaining the plan, conduct drills and exercises, and incorporate lessons learnt in the overall planning scheme.

5.3 Other structural and nonstructural measures

Long-term earthquake and tsunami mitigation measures, other than effective early warning system and efficient evacuation plan, are also important to reduce the damage caused on the shores by expected events. There is a wide range of both structural and nonstructural measures, implemented as pre-disaster mitigations to manage earthquake and tsunami risks in the coastal areas of Pakistan.

Within the framework of a coastal area management plan, measures which mitigate the impact of earthquake and tsunami hazard represent a coherent set of interventions. A project monitoring and control system should also be incorporated within such a plan.

This section describes the management of the earthquake and tsunami risk by strategic mitigation, both through the use of structural methods, including the use of natural coastal resources and engineering approaches, and also by nonstructural initiatives, including regulation and land-use, emergency response planning, and community preparedness.

5.3.1 Coastal engineering solutions

While it is not possible to prevent a tsunami, particularly in tsunami-prone countries, some measures have been implemented and tested to reduce the damage caused on the shores that may have succeeded in slowing down and moderating the impacts of tsunami, for example, construing tsunami in front of populated coastal area, raising ground level for housing near beach by infilling land, and building floodgates and channels to redirect the water from incoming tsunamis. However, their effectiveness has been questioned, as tsunamis are observed there, often higher than the barriers. These engineering options for risk reduction are analyzed, that is, if these are appropriate to the scale of the tsunami threat to the designated coastal area, balancing social and economic pressures against environmental considerations including sustainability, over the long-term.

It is important to have a reliable database of the building stock in coastal areas especially in close proximity of the shoreline to have a reliable vulnerability and thus risk assessment. Strong buildings, safe structures, and prudent land-use policies to save lives and reduce property damage are implemented as pre-disaster mitigation measures.

5.3.2 Hazard-resilient built environment

i. Relevant disaster management organization and building control authorities should maintain a GIS-based inventory of already constructed buildings with details of construction type, height, use, age, and structural stability at least for the areas in close proximity to shoreline (as a thumb rule, initially this database can be worked out for 1–3-km-wide coastal belt).

ii. Building control authorities in collaboration with other stakeholders should review and suggest policies to counter underlying challenges in the development of disaster resilient built environment. For example, lack of regulatory frameworks, unplanned cities and urbanization, old building stocks and at-risk infrastructure, unauthorized structures, weak institutional arrangements, inadequate capacities of local administration, lack of funding, inadequacy of qualified human resources, corruption, and unlawful activities are major challenges in this regard [6].

iii. Although cost may be an impediment, the national/provincial/local authorities should choose to adopt tsunami-resistant structures, stronger buildings, and deeper shock-resistant foundations mandatory in areas of high risk. The orientation of buildings with respect to the ocean is another factor for consideration. Mandated organizations should develop guidelines, byelaws, regulations, and codes to encourage coastal earthquake and tsunami-resistant infrastructure and housing in a local context. The overall general design guidelines could be developed from the experience gained from post-tsunami impact and damage assessments from different parts of the world as good practices.

5.3.2.1 Retrofitting of critical public buildings/facilities

i. Important public buildings, for example, schools, hospitals, and government offices, and infrastructure like telecommunication, communication, water supply, roads, and bridges shall be inspected by the concerned authority for evaluation against estimated earthquake and tsunami impacts. These evaluations shall contribute to the vulnerability and capacity assessments mentioned in Section 4 and completed on priority especially for facilities located within 3-km-wide coastal belt on priority, by the concerned government authority/organization.

ii. A complete framework for strengthening and retrofitting to be prepared and initiated according to available resources and fund generated through public-private partnership schemes.

5.3.2.2 Land-use planning

i. Information contained in the inundation, vulnerability, and risk maps is to be used as basis to develop policy on land-use planning for new development and to suggest critical measures to make existing land-use better resilient for fast-approaching tsunami [1].

ii. Hazard maps, particularly inundation maps for tsunami scenarios, are appropriate tools to suggest appropriate measure for land-use planning at any particular location. For example, option of "retreat" with reference to expected inundation extents can be used in land-use planning of high-risk areas [1].

iii. Coastal communities in Pakistan, especially in rural settings, tend to settle right on the beach without any appropriate set back distance from the shoreline. Thus, making these settlements much more vulnerable to coastal hazard at one end and an environmental hazard on the other, by throwing sewerage and garbage disposal directly into sea. Such settled remotely in small islands within the Indus creek system, these dotted communities are impossible to be warned in case of approaching local tsunami [5]. Relevant authorities and organization shall establish appropriate land-use management system to ensure a coastal hazard-resilient and environmentally sensitive land-use pattern along the coastline.

5.3.3 Strengthening of natural safeguards

Natural coastal features like high lands, sand dunes, mangroves, and other plantation species have been reportedly protecting the nearby communities in disaster situations. For example, interviews of 1945 Makran Tsunami survivors identified Pasni sand dunes, mangroves in Kalmat village, and Indus Creek system, mountain "Koh-e-Batil" at Gwadar, and tens of feet high rocks at Peshukan and Ganz villages as "savior" [7].

i. Coastal vegetation can be used to dissipate tsunami energy via turbulent flow through the media. The effectiveness of dissipation is dependent on the density of vegetation, its overall porosity, and its tortuous characteristics of porous matrix. It is important to consider that the vegetation itself is resilient against tsunami propagation and has a root structure that can resist the high velocity regime at the floor bed. Planting mangrove at appropriate locations can also serve to dissipate extreme wind wave energy.

ii. Sand dunes can provide natural full barriers against tsunami inundation. When overtopped, sand dunes tend to fail progressively by erosion. Dune-cladding vegetation provides reinforcement to the dunes, thus impeding erosion.

iii. Engineering solution for protection of coastal communities such as offshore breakwaters, dykes, and revetments can be used in hybrid way, i.e., with natural features, harnessing the full potential of coastal ecosystems including coral reefs, sand dunes, and coastal vegetation such as mangrove forests.

iv. Coastal development authorities, forest department, building control authorities, and local and provincial governments need to maintain a database (preferably GIS based) of such natural safeguard features in coastal belt and develop guidelines and regulation to protect and strengthen such features.

5.3.4 Continuity of operation plans for ports and other major facilities

i. Authorities of ports, large- and medium-scale industrial setups, and major facilities, including hospitals, schools, etc., both on- and offshore, need to prepare respective continuity of operation plans in response to estimated earthquake and tsunami impacts mentioned in Sections 3 and 4.

ii. The plan can include [8]:

• Conducting business impact analysis.

• Identifying recovery time objectives for business processes.

• Identifying recovery point objective for restoration.

- Define business continuity strategies and requirements.

- Work out procedures, resource requirements, and logistics for execution of all recovery strategies.

- Describing detailed procedures, resource requirements, and logistics for relocation to alternate work sites.

- Deciding detailed procedures, resource requirements, and data restoration plan for the recovery of information technology (networks and required connectivity, servers, desktop/laptops, wireless devices, applications, and data).

iii. National security agencies and port authorities also pay high attention to develop continuity of operation plan.

5.3.5 Debris clearance and management plan

Tsunamis of even small wave heights can bring huge quantities of debris and waste on the coast. Severe public sanitation and environmental concerns are also associated with earthquake and tsunami debris clearance and the management of municipal solid waste:

i. District government and municipal committees to develop tsunami debris clearance and waste recovery plan for expected tsunami derived waste estimation made either through numerical modeling or national and international case studies.

ii. National security agencies' infrastructure and ports need high-level consideration and debris clearance plan.

5.3.6 Emergency response, search, and rescue plan

i. All agencies and organization including civil defense, fire brigade, Rescue 1122, NDMA, PDMAs, Pakistan Army, Navy, Pakistan Coast Guards, Marine Security Agency, and health department mandated and/or capable of emergency response and rescue operations (even only in case of any critical situation) shall develop or adapt (already available) plans, SOPs, manuals, and guides as per estimated hazard of earthquake and tsunami with reference to Sections 3 and 4.

ii. The abovementioned plan and procedures shall be developed considering lead time availability of only "minutes" before a tsunami can hit the coast. To efficiently act upon the plans, strong coordination (inter and intra department) is to be assured through practicing envisaged plans and participating in scheduled drills and simulations coordinated by the NDMA and PDMAs.

5.4 Community preparedness

5.4.1 Database of tsunami knowledge

Tsunamis being infrequent phenomena could have gained least focus of all stakeholders in Pakistan; however, mega events of 2004 Indian Ocean tsunami and

2011 Japan tsunami created a sense of realization among national- to local-level organizations and experts to work on tsunami risk assessment and preparedness measures.

Within little more than a decade's period, significant pilot initiatives on community preparedness have been implemented based upon adaptation of international knowledge products and "Information, Education and Communication" (IEC) material. The adaptation strategy included not only the interpretation of those IEC products in national and local languages but also inclusion of indigenous knowledge and social and cultural traces. Tsunami safety tips, guidance for evacuation, observing natural signs of tsunami, protection, and conservation of natural safeguards of coastal region are delimited in handouts, pamphlets, information boards, booklets, videos, and through radio programs.

  i. The NDMA in collaboration with PDMA Baluchistan and Sindh need to maintain a database of all available IEC material for earthquake and tsunami community preparedness and education.

  ii. The NDMA to support PDMAs in finalizing standardization of available knowledge products and further adaptation including translation into local languages.

5.4.2 Public awareness campaign

  i. At local level, DDMAs need to plan and conduct tsunami awareness campaign on yearly basis through training of various community groups (volunteers, teachers, medical staff, local elected representatives, students, women, etc.) and of every Union Councils (UCs).

  ii. At national level, NDMA needs to design and implement public awareness campaigns focusing on earthquake and tsunami through national electronic channel and local FM radio channels using the tsunami knowledge database in the coastal region. The NDMA can collaborate with the Pakistan Electronic Media Regulatory Authority (PEMRA) to broadcast such information on electronic media channels under Section 20 (e) "Terms and Condition of License" of PEMRA Ordinance [9].

  "Broadcast, if permissible under the terms of its license, programmes in the public interest specified by the Federal Government or the Authority in the manner indicated by the Government or, as the case may be, the Authority, provided that the duration of such mandatory programmes do not exceed ten percent of the total duration of broadcast or operation by a station in twenty-four hours except if, by its own volition, a station chooses to broadcast such content for a longer duration."

  iii. The NDMA in collaboration with the Pakistan Telecommunication Authority (PTA) can be vital to design and implement tsunami awareness campaigns for general public in coastal region through social media and cellular phone networks.

  iv. The NDMA, in collaboration with PDMAs and other government and nongovernment organization, can manage (if already available) and develop (if not readily available) knowledge products on a standardized format for public awareness campaign on the following subjects:

- Observing natural signs of tsunami

- Receiving and responding official warnings

- Identification and recognizing evacuation centers and routes

- Information about hazard-resistant construction, land-use, byelaws, regulations and codes to ensure safety against earthquake, tsunami, flooding, and fire impacts

- Evacuation procedures and guidelines

- Importance of participating in evacuation drills and training

- Conservation and strengthening of natural safeguards of tsunami like mangroves, high land, sand dunes, and coral reef.

The information mentioned above shall be used through electronic, print, and social media campaigns and community training.

v. Evacuation drills must be conducted to ensure training of the community on disciplined evacuation. A regular schedule of conducting drill shall be planned and implemented at local level (UC and village level) once a year with communities settled on the coastline (at least within 3-km-wide coastal belt). PDMAs (Baluchistan and Sindh) shall support respective DDMA to design and implement community-led and sustainable mechanism for monitoring entire processes.

5.4.3 Curriculum development for all levels of academia

i. The NDMA in collaboration with PDMAs should lead the process of finalizing curriculum on earthquake, tsunami, flood, cyclone, and fire hazards and preparedness measures for students of all levels. Education departments shall be a part of this process. Private educational institutions also follow the finalized curricula.

ii. Schedule of evacuation drill (at least once a year) in all level academic institutions, both public and private, located in coastal districts shall be finalized by respective DDMA and district education department.

iii. Teachers' training program to be developed by PDMA Baluchistan and Sindh in collaboration with provincial and coastal districts' education departments.

5.4.4 Self-evacuation plan for remote coastal communities

Fishing villages in coastal Pakistan along tidal creeks of the Indus Delta and Makran region would need to respond quickly to escape a tsunami from nearby parts of the Makran subduction zone.

i. The NDMA in collaboration with PDMA Baluchistan and Sindh should conduct a survey and mapping (using GIS) of all remote coastal communities where means of official warning communication are limited or not at all available, in estimated available lead time (subject to the estimated tsunami hazard and risks discussed in Sections 3 and 4).

ii. PDMAs (Baluchistan and Sindh) along with respective DDMAs should design and implement evacuation planning for such communities including:

    a. Identification of feasible evacuation site and routes near each individual settlement.

    b. Awareness campaigns and training of local volunteers to receive official warning to disseminate to other villagers and fishermen.

    c. Interpretation of bulletins issued by the PMD and DDMAs.

    d. Detection of early warning via natural signs such as abnormal behavior of animals, earthquake shaking, and retreat of sea water.

    e. Basic emergency response, especially how and where to evacuate, immediate first aid provision to injured, etc.

    f. Facilitate and mange evacuation of vulnerable groups.

    g. Ways to manage external communication to get help from outside of the village and emergency responders/organization.

    h. Knowing, using, and keeping alive the indigenous knowledge.

    i. Knowledge about different categories of threat and how they should respond to it.

5.5 Risk transfer

The NDMA in collaboration with PDMAs (Baluchistan and Sindh) and other organizations should develop a strategy to promote insurance (life and property) for earthquake and tsunami incidents that can play an important role in offering financial protection from the costs of disaster.

5.6 Research and knowledge sharing

Understanding disasters and to find appropriate ways to reduce disaster risk are critically important. Scientific, social, and indigenous knowledge-based researches are direly needed to be undertaken, and result sharing with larger audience including communities at risk has a pivotal role in managing disasters. This role of risk-based knowledge sharing has been recognized in international frameworks, i.e., Sendai Framework for disaster risk reduction (SFDRR) 2015–2030 [10].

i. At national level the NDMA may facilitate coordination among academia, research institutions, and private sector to undertake scientific and social research initiatives facilitating overall risk assessment of coastal areas of Pakistan on:

    a. Earthquake and tsunami hazard analysis

    b. Exposure data, vulnerability and capacity evaluations

    c. Indigenous knowledge

    d. Preparedness and response

    e. Related policies, regulations, guidelines, bylaws, and codes

  ii. The Higher Education Commission (HEC), the National Institute of Oceanography (NIO), PMD, NDMA, PDMA Sindh, and PDMA Baluchistan should explore public-private partnership to encourage researchers to undertake the required studies mentioned in this section as well as ensure that the results are shared and available to the end user (including general public) and are incorporated in policies, regulations, and guidelines from national to local level.

5.7 International cooperation and coordination

International cooperation on tsunami warning and mitigation is envisaged to assure international compatibility and interoperability for rapid exchange of data and information. Pakistan is actively engaged in exchange of data and resources and capacity building initiatives through bilateral and global commitments. Pakistan is a member state of the Intergovernmental Oceanographic Commission of UNESCO (IOC-UNESCO), established in 1960 as a body with functional autonomy. The Pakistan Meteorological Department is the focal agency mandated to coordinate with UNESCO-IOC, ocean-wide tsunami warning providers for data sharing and capacity building regional and global initiatives.

1. The PMD in coordination with disaster management authorities at national and provincial level should ensure to participate in all capacity building initiatives and ocean-wide simulations and drills.

2. Effective participation in research, knowledge sharing, and capacity building should be ensured, and the PMD being the focal agency should play a lead role.

3. The PMD should also play a lead role in appropriate follow-up of global and regional collaborations from national to local level.

6. Conclusion

Coastal area residents in Baluchistan and Sindh provinces can experience a local earthquake—the most common cause of tsunamis—and a local tsunami generated in Arabian Sea can approach the coast within minutes.

Limited information regarding Pakistan coastline's vulnerability is available to assess tsunami risk. Database is not appropriately maintained for social, physical (structural), economic, and environmental dimensions of exposure analysis, making the situation more critical. Since 2006 (in the after math of 2004 Indian Ocean Tsunami), some limited but focused efforts on tsunami hazard and risk assessment, mitigation, and preparedness have been piloted in the country since 2006 serious and consistent efforts of all stakeholders at policy and implementation level.

This chapter suggests earthquake and tsunami risk assessment and mitigation roadmap for Pakistan's coastal areas with a vision of acquiring required deposit of information to plan and implement coherent and synergized earthquake and tsunami risk reduction measures. The guidelines for in various sections of the chapter are proposed for:

i. Determining the earthquake and subsequent local tsunami threat, in terms of hazard and risk assessment all along the coastline on a standardized pattern on priority basis, involving all levels of stakeholders.

ii. Ensuing federal and provincial agencies utilize earthquake monitoring systems, tide gauges, deep ocean buoys, and other capabilities (international/regional information sharing systems to gather as much information as possible about a potential tsunami). Essential data is then used by forecasting and analysis centers for the assessment of the immediate tsunami threat. Timely and accurate warnings must then be disseminated in clear and actionable terms to emergency managers and a ready public.

iii. Identifying mitigation strategies that involve sustained actions taken to reduce or eliminate the long-term risk to human life and property based on earthquake and tsunami risk assessments.

iv. Aiming at tsunami-resilient communities that have plans, enhanced communications, and heightened awareness of the citizens to ensure resilience to earthquake and tsunami events, reduced economic losses, and shorten recovery periods.

v. Encouraging continued broad scientific and social research efforts needed to improve all-purpose understanding of tsunami processes and impacts and then to develop more efficient and effective risk assessment, risk communication, prediction and warning, preparedness, and mitigation measures.

vi. Strengthening partnerships with international organizations and other countries persuade bilateral and multilateral agreements to better understand and reduce the common threat and impact of earthquake and tsunami in the region.

Author details

Ghazala Naeem
Resilience Group, Islamabad, Pakistan

*Address all correspondence to: ghazala_ghq@hotmail.com

IntechOpen

References

[1] IOC Manuals & Guide 52. Tsunami Risk Assessment & Mitigation for the Indian Ocean. Intergovernmental Oceanographic Commission. 2009. pp. 16-17, 33, 45, 58-69. Available at: http://iotic.ioc-unesco.org/images/xplod/resources/material/tsunami_risk_270809_lr.pdf

[2] Heidarzadeh M. Tsunami Risk, Preparedness and Warning System in Pakistan: Chapter 6. In: Atta-Ur-Rahman editor. The Book Disaster Risk Reduction Approaches in Pakistan. Springer; 2015. p. 120. DOI: 10.1007/978-4-431-55369-4_6

[3] Hoffman et al. 2009: An Indian Ocean Tsunami triggered remotely by an onshore earthquake in Baluchistan Pakistan. Geology. 2014;4. DOI: 10.1130/G35756.1

[4] Multi-Hazard Risk and Vulnerability Assessment Guidelines (MHRVA). National Disaster Management Authority Pakistan; 2016. pp. 44-48, 75-91. Available from: http://ndma.gov.pk/publications/MHVRA%202017.pdf

[5] Naeem G, Nawaz J. Challenges and opportunities for reducing losses to fast arriving tsunami in remote villages along the coast of Pakistan. In: Mukhtari M, editor. Tsunami. Intech; 2016. pp. 136-163. ISBN 978-953-51-2676-8

[6] Malalgoda C et al. Challenges in Creating a Disaster Resilient Built Environment. 2014. pp. 139-144. DOI: 10.1016/S2212-5671(14)00997-6

[7] Remembering 1945 Makran Tsunami booklet: United Nations Educational, Scientific and Cultural Organization (UNESCO) Intergovernmental Oceanographic Commission (IOC). 2015. pp. 17-64. Available from: http://iotic.ioc-unesco.org/1945makrantsunami/1945-makran-tsunami-booklet.pdf

[8] FEMA Business Continuity Plan. 2014. Available from: https://www.fema.gov/media-library-data/1389019980859-b64364cba1442b96dc4f4ad675f552e4/Business_ContinuityPlan_2014.pdf

[9] Pakistan Electronic Media Regulatory Authority (PEMRA) Ordinance 2002; Terms and Conditions of License. p. 18. Available from: http://58.65.182.183/pemra/pemgov/wp-content/uploads/2015/08/Ordinance_2002.pdf

[10] Sendai Framework of Disaster Risk Reduction (2015-30); Priority 1. Available from: https://www.unisdr.org/we/coordinate/sendai-framework

Section 3

Health Consequences

Outbreak of Traumatic Defeat Earthquakes: Health Consequences and Medical Provision of the Population

Diana Dimitrova

Abstract

Earthquakes are described as the most destructive and unpredictable disasters around the world. Many types of consequences are presented as possible negative effects including health-related of them. Identifying and categorizing the various health risk factors is an initial goal in an uncomplicated earthquake setting. Effective organization of the health care system (HCS) in case of complicated medical situation due to earthquakes and tsunamis is a serious challenge. The healthcare system operates at high speed with considerable difficulties in the event of a large magnitude outbreak of a traumatic defeat an earthquake. Possibilities to take adequate solutions in conditions of the worst-case earthquake scenario with the subsequent provoked multi-secondary disasters as tsunamis and with multi-secondary risk factors are highly motivating for the medical community with critically low resource constraints. On the one hand the analysis of the structure of mass victim and medical triage in a complicated scene due to earthquakes is a difficult process. On the other hand medical provision of the population in highly destructive earthquakes is limited by time.

Keywords: earthquakes, health consequences, medical provision of the population, emergency and disaster medicine

1. Introduction

In case of emergencies and disastrous situations (EDS) the medical provision of the injured population (MPIP) is a key point in the emergency plan of the health system (HS) [1–3]. This is strongly regulated by the HS of the country [2]. The action plan of EDS is a consistent, up-to-date annual task in order to maintain a highly optimal readiness for rapid response [2–4]. The availability of action protocols for each individual critical care potion guarantees confident firmness and readiness for an operational response [5–7]. Medical provision of the population with sufficient resources is the basis step to solve during managing the situation [7, 8]. Interaction with other systems is essential for rescuing and providing effective medical care to the affected population [8, 9]. The coordination of the EDS activities on a large scale is a solid fundament principle for the correct direction of the actions [7–9]. The participation of international organizations, forces and resources is a possible option in case of the EDS with a serious territorial scope and severely affected available resources [10].

A special type of disasters, which is characterized not only by mass but also by the diversity of damage to the population are high magnitude earthquakes and tsunamis [2, 7–9].

According to the World Health Organization (WHO), on average more than a million tremors occur annually in the world, of which about 100,000 have a magnitude of 3–8 Richter and are felt by humans. Some of the strongest earthquakes in the world are: the Assam (June 12, 1897) in Northeast India; The Japanese (September 1923), in which the cities of Tokyo and Yokohama were destroyed; Gobi - Altai (December 4, 1957); Chilean (May 29, 1960); the Alaska Earthquake (March 28, 1964); the Armenian Earthquake (December 7, 1988); the earthquake and tsunami of December 26, 2004 after which almost 230,000 missing and presumed dead [11]; the earthquake in Haiti (January 12, 2010) that killed more than 230,000 people and another 300,000 were injured [12]; the earthquake and tsunami in Japan (March 11, 2011), which killed more than 15,800 people, injured more than 6000 and disappeared more than 2500 people [13].

2. Medical provision of the population: Goals, principles and tasks

The medical provision of the population (MPP) is an element of an activity and a plan of the health care system especially due to EDS [7]. In case of emergencies it is based as much as possible on the existing health care system and only with an organizational approach moves to a new mode of work with available staff and sources [8–10]. MPP according to real practice, results of an epidemiological survey and documentary research is defined as a complex of interconnected organizational, medical and hygienic-anti-epidemic measures [7–10, 14].

It seems that the Aim of MPP is organized in a few groups of actions:

1. *Preservation of the health* and strengthening of the physical condition and working capacity of the population.

2. *Saving the lives* of the affected people and reducing mortality and disability and the fastest recovery of health and work ability.

3. Prevention of long-term diseases.

4. *Preventing infection diseases*.

The occurrence of different types of traumatic defeat due to huge EDS with diverse injuries [7–10, 14, 15] determines the main principles of MPP [7–10].

The main Principles of MPP due to contemporary scientific knowledge are collected in some main key points [7–10, 16]

1. *Universality* of using of medical resources.

2. *Maximum allowable economy* of using of available resources.

3. *Implementation of the medical evacuation* into a MPP activity.

4. To *use correctly* Unified rescue system with unified emergency number for providing MPP.

5. To use *unified doctrine* for rendering medical aid and treatment from the epicenter of defeat to the ends in the multi-profile and specialized medical establishments until the final outcome.

For the correct understanding and optimization of the activities connected with EDS the tasks of the MPP are divided into three groups depending on the time for their implementation – before, during and after disaster strike [7–10, 16–18].

1. Before:

   - Study of the devastating effect of various factors in EDS and the means for prevention, diagnosis and treatment.

   - Planning the activities of the health system for MPP.

   - Construction and maintenance of formations and resources of the MPP.

   - Creation of the medical and sanitary property.

   - Training, drills, workshops for preparing medical staff to react as better as possible.

   - Constant and regular current hygienic control.

2. In case of threatening and an ongoing EDS:

   - Deployment of the medical formations according to the operational plan, keeping them ready for work and protection.

   - Bringing the health risk management system into readiness.

   - Strengthening the medical provision (MP) system and teams and start MP of the evacuated population from epicenter, during transport to emergency room.

   - Strengthening and targeting the operational and special preparation of medical staff to starting and working into epicenter.

   - Strengthening the epidemiological surveillance of the territory.

3. After:

   - Medical reconnaissance.

   - Introduction of the formations in the center of defeat.

   - Organization and rendering the first medical aid in the center of defeat.

   - Removal and evacuation of the victims for appropriate care and subsequent treatment.

   - Provision of complex of hygienic and anti-epidemic measures.

   - Continuous management and maneuver for the most appropriate use of the medical forces and means.

   - Conducting a forensic medical examination of the victims and their identification.

Proper performance of these tasks is a prerequisite for providing in the shortest possible time the optimal amount of medical care to the largest possible number of victims [2, 7–11, 14, 16–18].

3. Earthquakes as an outbreak of traumatic damage: Risks and consequences

3.1 Mass destruction after earthquakes

Affecting large areas with mass destruction and mass loss is a typical effect after hug magnitude earthquakes [2, 7–10, 13, 19–22]. A typical example of possible consequences in large-magnitude earthquakes is presented in **Figures 1** and 2 in Bulgaria due to some of the most destructive earthquakes in XX century (**Figures 1** and 2) In Gorna Oryahovitsa region, M = 7, 1913 and in Blagoevgrad region, M-7,8, 1904 are described the biggest earthquakes in Bulgaria. Totally are destroyed respectively Gorna Oryahovitsa and Kresna. Industrial sites, homes, hospitals, public buildings, utilities, underground and aboveground technical facilities, transport hubs, etc., are damaged or destroyed [2, 7–10, 13, 19–22]. This significantly impedes rescue operations, effective enough immediately after the earthquake occurs [7–10].

3.2 Secondary disasters due to earthquakes

Analysis of primary and secondary statistical data [2, 19–21, 23–26] shows that occurrence of secondary defeats after an earthquake is a common effect [7–10]. This increases number of risks factors and character of injuries among the affected population. According to analysis some of possible secondary disasters after earthquakes are [2, 7–10, 13, 19–21, 23–26]:

1. *An outbreak of chemical contamination* by industrial poisons. Observed in the area of the incident due to the destruction of chemical enterprises, warehouses with agricultural or industrial poisons. The chemicals produce not only acute intoxication but also long-term heath effect and cancer diseases as well [7, 10, 19].

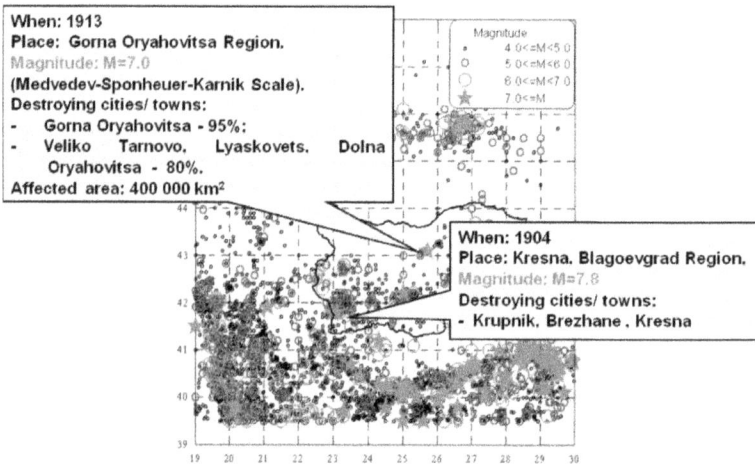

Figure 1.
Large-magnitude earthquakes in Bulgaria-in Kresna and in Gorna Oryahovitsa. Evidence-base practice database (XX century) (Sources: National Institute of Geophysics, Geodesy and Geography (NIGGG) in Bulgaria.)

Figure 2.
Large-magnitude earthquakes in Bulgaria-in Kresna and in Gorna Oryahovitsa. Photos (XX century)
(Sources: National Institute of Geophysics, Geodesy and Geography (NIGGG) in Bulgaria).

2. Occurrence *of explosions and fires* due to destruction of flammable or explosive objects as petrol station after large magnitude earthquakes. They can develop dangerous additional consequences as fire burn, chemical burn, traumas, toxicological injuries etc [7, 10].

3. Destruction of buildings and severe *air pollution* with fine dust particles in large enough areas because of an earthquake leading to casualties, poisoning and suffocation [7, 10, 13, 19, 23–26].

4. *An outbreak of nuclear damage* such as Fukushima Daiichi nuclear accident, 2011 due to the large earthquake and tsunami is a real consequences. Radiation as danger risk factor produce particular health problems, which can be: acute, or chronic; somatic or genetic; stochastic and non-stochastic diseases; and long-term effects [7, 10, 13, 23–26].

5. *A tsunami* is generated often by underwater earthquakes. Fukushima 2011 and Haiti 2010 are significant examples and it is shows that many health problems influence not only primary effected county, but also much more people around the world for many years [1, 7, 8, 13].

6. Occurrence of *catastrophic floods* [1, 7–9, 11–13] after earthquakes can increase number of traumas due to new risk factors with leading place and role for the speed and depth of the water flow. On the other hand can cause hypothermia [7, 10, 27] especially for infants, adults, for the chronically ill and for people with special needs like people with compromised vision or limbs. Can be formed a new defeat of waterborne disease such as Cholera as well (like Haiti, 2010) [1, 7–9, 12].

7. *An outbreak of biological contamination* and infectious diseases can be created after a destructive earthquake within or without a tsunami wave (Like Haiti, 2010; like Nepal, 2015) [1, 7–9, 21, 22].

8. Occurrence *of landslides and avalanches* (Nepal, 2015) [1, 21, 22].

9. *An ecological disaster* and even local environment can be changed totally (Kresna, Bulgaria, April, 1904, with 7,8 M) [7–9].

10. *A volcano* generated by an earthquake as a secondary disaster. Discarded volcanic ash and toxic substances affect the environment and people [7–9].

11. *Traffic accidents* due to earthquakes [7–10].

12. *Social related and financial crisis* due to earthquakes [1, 7–10, 13, 17].

3.3 Victims: Main structure and frequency of traumatic health related effects

It seems that after huge earthquakes a large number of victims and injuries are occurred. According to experts, many numbers of victims need emergency medical care at the same time [7–10]. The nature and structure of the injuries can be very diverse [7, 13]. Injuries to the musculoskeletal system, extensive burns, prolonged compression syndrome due to prolonged compression of individual limbs or parts of the body, injuries to large blood vessels predominate [7–10]. A significant percentage of victims may be in a state of shock [7, 10, 27], with acute respiratory and cardiovascular insufficiency, in need of urgent respiratory and cardiovascular resuscitation [7, 10], neuropsychiatric disorder due to the experienced mental stress and others. According to scientific research [7, 10], by severity, medical losses are divided into the following groups: lightly injured (40%), moderately severely and severely injured (60%), of which 20% need specialized medical care.

3.4 Pollution from different origin as health risk factor

Creating a severe hygienic-epidemiological situation on the territory of the outbreak of traumatic defeat is defined as a big health risk factor. Prerequisite for this is the pollution of the territory from the destruction of water supply and sewerage systems, difficulties in finding the corpses of dead people and animals, the appearance of rodents, insects etc. The appearance of diseases of infectious and non-infectious origin is possible, as well as the appearance of epidemics - typhoid fever, paratyphoid fever A and B, salmonellosis, hepatitis, cholera etc. [7–10].

4. Health risk management: Essential principles

4.1 Role and place of the medical forces and resources for MPIP

4.1.1 Emergency care system (ECS)

Usually the signals for victims of an accident or disaster are received by the medical director (manager) of the emergency room and emergency hospitals as medical institution from ECS. Firstly, emergency medical care center (EMCC) as a front line of health system is informed by Unified rescue and emergency number 112 (UREN-112). (**Figures 3** and **4**) A general scheme of UREN-112 system in Bulgaria is presented in **Figures 3** and **4**. By National System – 112 among affected area, start working together teams of EMCC and Fire Safety and Civil Protection directorate (FSCPD). Police as a part of Ministry of Interior (MI) take place and

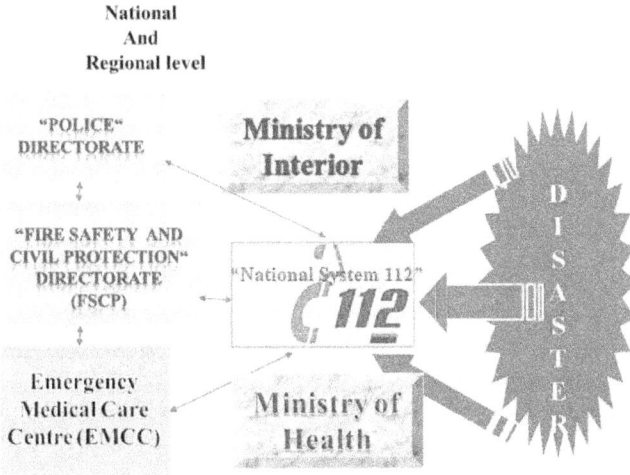

Figure 3.
A general scheme of UREN-112 in Bulgaria.

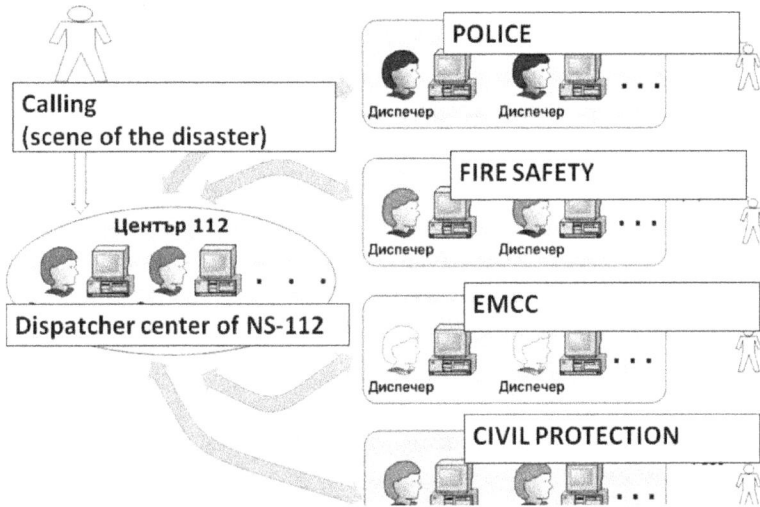

Figure 4.
A general scheme of NS-112 in Bulgaria.

control the entrance and exit of the affected area. The teams of EMCC are the first to go to the scene of the EDS as disaster event [7–10, 14, 16–18, 27].

In large-scale disasters, the EMCC teams are not enough to provide the necessary amount of medical care to those in need. This requires in advance being formed, prepared and equipped additional teams of staff of medical institutions for their inclusion in the provision of medical care. In addition to these medical teams for the population, especially in earthquake-prone areas or areas with chemical sites in anticipation of numerous medical losses, emergency hospital medical teams and emergency military teams can be used. These are medical formations built on a functional principle with opportunities to provide emergency qualified therapeutic and surgical care for vital indications [7, 14, 16–18, 27].

4.1.2 Hospitals

The main tasks of the hospitals in EDS are the provision of medical care and treatment of the victims and hygienic and anti-epidemic provision of the affected regions, and by order of the chairman of the respective commission (district, municipal) and neighboring regions [7].

The network of health facilities and their infrastructure must be ready to provide timely emergency and specialized medical care to the population in emergency and disaster situations [2, 7–9, 16–18, 27].

Knowledge of the factors that can lead to damage to health or endanger the lives of people in EDS allows them to predict the medical consequences, to clarify ways to combat them, to take the necessary preventive measures to limit the medical consequences, to organizing emergency medical measures and eliminating the consequences of emergencies [7, 8].

The hospitals provide the necessary human and material-technical resources, create an effective organization and keep in constant readiness the forces for immediate action in EDS [2, 8].

Before the occurrence of the disaster, the head of the medical institution – hospital must make a comprehensive assessment of the condition and the ability of the health institution to work in such a situation. During this period, the action plan for the EDS and the work of the medical institution for the medical provision of the population of the respective territorial unit must be developed, in accordance with the plan of the Ministry of Health [2, 8, 16–18, 27].

The plan is developed in different variants depending on the expected nature and severity of the medical losses and includes the following [7, 8]:

1. Creation and maintenance of a system for notification of the employees of the hospital.

2. Calculation of the medical losses and of the necessary forces and means for rendering medical aid to the victims.

3. Necessary medical teams and formations for rendering emergency and urgent medical aid, as well as inpatient medical care.

4. Creation of an appropriate structure (restructuring) of the bed base of the medical institution if necessary.

5. Organizational scheme for providing medical care at the site of the lesion, during transportation and in the medical institution.

4.1.3 Regional health inspectorates (RHI)

The Regional Health Inspectorates (RHI) are developing a work plan for the Hygiene and Epidemiological Provision and Inspectorate in the disaster area. It must be in accordance with the plan for conducting rescue and other urgent works of the regional and municipal commission [7, 8].

In case of disasters, the director of RHI [7–9] clarifies the place and nature of the event, then organizes and conducts research and control of environmental hygiene parameters in the affected areas, in industrial and other sites in terms of toxic substances, dust, noise, vibration, microclimate, radiation and other harmful factors. This activity is carried out by pre-formed and trained anti-epidemic teams of RHI on the territory of the disaster [7].

The main task of the RHI is the organization and implementation of disinfection, disinsection, deratization and control of the degassing and decontamination activities in the affected areas after the normalization of the situation [8].

The number and nature of foodstuffs affected by the disaster, the type and quantity and the nature of the damage must be clarified, and enhanced sanitary control must be organized over all foodstuffs in the disaster area. This requires organizing and conducting intensive laboratory control over the affected catering establishments and food industry establishments.

Based on the conclusions of the analysis, the RHI prescribes measures for compliance with hygiene standards and requirements for all factors of the working environment. After conducting a control for hygienic efficiency of the conducted measures, a conclusion is given for safe working conditions with a view to resuming regular operation of the affected sites [7, 9].

4.2 Organization of MPIP

4.2.1 Surgical and trauma care

The relevant clinics from the Multidisciplinary Hospital for Active Treatment (MPHAT) and Hospital for Emergency Medicine, as well as the clinical departments of surgery, orthopedics and traumatology, resuscitation and anesthesiology in the district, regional and municipal hospitals are used as a base. If necessary, the bed capacity of the same hospitals is used. In some cases, staff and facilities from other surgical units (ophthalmology, maxillofacial surgery, etc.) of hospitals can be used. This allows in EDS for a short time and without significant difficulties to be included in the organizational scheme of medical care. About 60% of the inpatients can be discharged and a specialized bed stock can be released for the needs of the victims. For work in a trauma center, if necessary, medical teams (trauma, surgical, etc.) are formed on a functional principle, without seriously violating the readiness of these wards for admission and treatment of victims. These teams must arrive at the scene no later than one hour after the emergency medical teams. At this time, the medical situation, the scope of work and the possible number of required specialized surgical teams should be clarified. If necessary, they can be strengthened with teams from medical institutions in neighboring regions [7, 8].

4.2.2 Radiological care

In case of accidents at the NPP, in case of incidents with sources of ionizing radiation, in case of cross-border transfer of radioactive substances in the therapeutic wards of the MPHAT, an opportunity must be created to provide radiological assistance to the victims. All therapeutic wards of the medical establishments, in the vicinity of the NPP, must be ready for possible admission of radiation patients and those with combined radiation injuries. For this purpose, it is necessary for physicians-therapists to have radiobiological training and in case of radiation conditions to organize the work of the ward in radiological terms and to conduct radio protective measures in the medical institution. The existing departments of radiotherapy and isotope diagnostics, based most often in oncology dispensaries, oncology hospitals, etc., have a corresponding place in this functional radiological system. The medical staff from these radiology departments are involved in providing radiological assistance to the victims [7, 20–26].

The duty and responsibilities according to the International Atomic Energy Agency (IAEA) require doctors to have the relevant knowledge of radiation protection, which enables them to initiate preliminary treatment and provide assistance to specialized units in the event of a radiation accident. Another task of health care in the section of radiological care is the control of the radiation parameters of the working and living environment, which directly affect the person [22]. The radiation control department must organize and conduct the necessary radiation-hygienic measures on the given territory.

The organization of the radiological assistance is related to the plan for radiation protection of the country in case of an accident at the NPP, which ensures the implementation of the plan in its medical section [7, 20–26].

4.2.3 Toxicological care

The organization of toxicological care uses a mixed approach, including the establishment of staff and functional units. The expanding chemical pathology necessitated the establishment of full-time clinics and toxicology departments in the settlements with large sites of the chemical industry. These units, in addition to providing toxicological assistance to the population, also serve to train medical personnel in this field. In the other hospitals, the therapeutic wards are re-profiled into toxico-therapeutic ones for admission of toxicologically ill patients for emergency toxicological care. Good interaction should be ensured with the intensive care unit of the hospital [7, 10].

On the basis of the staff clinics and toxicology departments, specialized medical teams are established, provided with medical and sanitary equipment and transport [7]. These teams must be constantly prepared to work in a chemical outbreak or to strengthen the therapeutic wards of neighboring hospitals where toxicologically ill patients are hospitalized [7–10].

4.2.4 In outbreak of biological contamination (OBC)

When creating an outbreak of biological infection all types of medical care (first medical, qualified and specialized) [7] are within the area of the outbreak. For this purpose, the medical and prophylactic establishments on the territory in the OBC mainly are used [7–10].

4.2.5 In the outbreak of combined defeat

The first medical and qualified medical care for vital indications is provided at the medical center according to various schemes [7]. Most often, two groups of medical forces are created for rescue operations in the center of a combined defeat: in the biologically infected area and outside it [8, 9].

5. Medical triage in an OTD

Medical sorting [7] is performed in the OTD. According to the severity of the OTD [7, 9], the distribution of the victims by sorting groups allows for homogeneous treatment and prevention measures [8, 10, 16, 17].

Depending on the severity of the injuries the victims are sorted into two main groups in Bulgaria [7, 10]:

    a. Slightly injured (40%). This group includes victims of soft tissue injuries who do not need hospital treatment;

b. Moderately and severely injured (60%). These are victims who need urgent medical attention and inpatient treatment. This group can be divided into four subgroups:

- Group T1. Persons with immediate vital disorders (20–40%). This group includes victims with respiratory failure, cardiac arrest, ventricular fibrillation, huge bleeding, shock, increased intracranial pressure, burns of the face and respiratory tract, or extensive burns occupying more than 20% of the body surface; poly-trauma. The victims of this group receive emergency first aid in order to stabilize the basic vital functions and have priority in treatment;

- Group T2. Persons to whom medical care can be postponed for 6–8 hours (20%). These victims have an advantage in transportation, but do not need extreme treatment. These include victims with some thoracic and abdominal injuries, injuries to the uro-genital tract and some blood vessels, burns less than 20% of the body surface;

- Group T3. Injured with some cranio-cerebral injuries, some spinal cord injuries and slightly injured (40%). These are persons with injuries of the small bones of the frontal part of the skull (mandible, nose, orbits), medium and small soft tissue injuries, etc.;

- Group T4. Dying and agonizing.

Particular attention in medical sorting should be paid to groups dangerous to others and in need of urgent medical attention. Dangerous for others are those infected with poisonous substances, radioactive substances, bacterial agents and patients with particularly dangerous infections (PDI), mental disorders etc. This danger imposes the need for sanitary treatment of the infected and isolation of patients with PDI, mental disorders as well.

6. Preventive measures for primary risk reduction

Prevention of the population in favor of public health and the future of the nation is fundament for stabilization of the health system in case of EDS [7, 8, 16, 17]. Prophylaxis measures (PM) used for primary risk reduction is only first group of tasks and primary step for government and national health system. **Figure 5** presents the main groups of measures for prevention of destructive earthquakes distributed in time: before, during and after. (**Figure 5**) According to science society some main PM are:

1. Earthquake forecasting.

2. Establishment of well-equipped and prepared seismic stations and a notification system.

3. Scientifically based anti-seismic construction.

4. Compilation and timely updating of a map of OTD threats.

5. The approaches to the hospitals should be known and organized in case the entrances are covered with destructions. For each hospital, which is located in an area with high seismicity, a helipad should be provided.

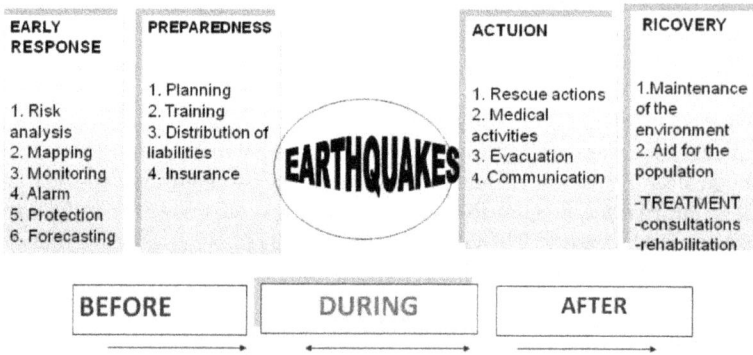

Figure 5.
The main groups of measures for prevention of destructive earthquakes distributed in time: before, during and after.

6. Systematic training of medical forces

7. Systematic preparation of the population to react and provide main first medical aid steps.

7. Conclusion

Earthquakes, unlike other disasters, are characterized by sudden onset and rapid flow. An in-depth analysis of earthquakes shows that they are characterized by a number of specific features. In earthquakes the response time is practically a very limited resource. The multi-factorial nature of the risks due to an earthquake and the possible consequences require good preparation and collaboration of various institutions with the healthcare system in the OTD. The medical provision of the affected population includes all levels of management of the health care system. The field work on the medical provision of the injured people is based on the EMCC. Proper medical triage and the provision of medical care in OTD is a staged process. In case of traumatic defeat of EDS it is use the two-stage system with evacuation by appointment as method for MPIP.

Conflict of interest

The author declares no conflict of interest.

Author details

Diana Dimitrova
Emergency CC, Department of EM and MCS, Stanke Dimitrov, Bulgaria

*Address all correspondence to: d.dimitrova.phd.md@abv.bg

IntechOpen

References

[1] The UN. Natural Hazards, Un Natural Disasters. The Economics of Effective Prevention 24-30, Available from: http://documents.worldbank.org/curated/en/620631468181478543/pdf/578600PUB0epi2101public10BOX353782B.pdf

[2] EC.JRC.IPSC. European Commission. Joint Research Centre. Institute for the Protection and Security of the Citizen. Current status and best practices for disaster los data recording in EU member states., Available from: http://pprdeast2.eu/wp-content/uploads/2015/11/JRC-SOTA-Best-Practices-Loss-Report.pdf

[3] WHO. Earthquakes, Available from: https://www.who.int/health-topics/earthquakes#tab=tab_1

[4] BGS. Earthquake database search, Available from: http://www.earthquakes.bgs.ac.uk/earthquakes/dataSearch.html

[5] NOAA. The Significant Earthquake Database, Available from: https://www.ngdc.noaa.gov/nndc/struts/form?t=101650&s=1&d=1

[6] CGS. Earthquake Data and Reports, Available from: https://www.conservation.ca.gov/cgs/earthquake-data

[7] Dimitrova D., Medical provision of the population during earthquakes - readiness of Emergency medical care center in Blagoevgrad region. [PhD thesis].Sofia: MU; 2015.

[8] Dimitrova D., Management of medical provision of the population during earthquakes in Bulgaria. Current state and development perspectives. 1st ed. Propeller: Sofia; 2016.

[9] Dimitrova D.: Challenges in emergency medical care center in Bulgaria in case of mass traumatism during disaster situations – Epidemiological survey, 2018; 130:e139. DOI:10.1016/j.resuscitation.2018.07.299

[10] Dimitrova D., Medicine of the disastrous situations. Educational exercises and seminars - from theory to practice. V1. 1st ed. Propeller: Sofia; 2016

[11] Lay, T.; Kanamori, H.; Ammon, C.; Nettles, M.; Ward, S.; Aster, R.; Beck, S.; Bilek, S.; Brudzinski, M.; Butler, R.; DeShon, H.; Ekström, G.; Satake, K.; Sipkin, S. (20 May 2005). "The Great Sumatra-Andaman Earthquake of 26 December 2004"(PDF). Science. 308 (5725): 1127-1133.

[12] The balance. Haiti Earthquake Facts, Its Damage, and Effects on the Economy. The 2010 Earthquake Caused Lasting Damage, Available from: https://www.thebalance.com/haiti-earthquake-facts-damage-effects-on-economy-3305660

[13] NPAJ. Damage Situation and Police Countermeasures.8 March 2019. National Police Agency of Japan. Retrieved 13 March 2019

[14] Dimitrova D., Evidence based medical practice. Aspects of emergency and disaster medicine. 1st ed. Propeller: Sofia; 2015.

[15] WHO. Mass casualty management systems: strategies and guidelines for building health sector capacity, Available from: https://www.who.int/publications/i/item/mass-casualty-management-systems-strategies-and-guidelines-for-building-health-sector-capacity

[16] Dimitrova D., Role and place of emergency medical care center in Republic of Bulgaria for providing pre-hospital care during disaster situations.

PDM, vol 28, supplement 1, Publ. CUP, 2013, ps97

[17] Dimitrova D., Psychiatric aid in earthquakes.-In: Medical and psychiatric provision in missions and operations abroad - contemporary challenges. S., 2012, p13

[18] Dimitrova D., Effective management of the working shift in EMCC [thesis]. Blagoevgrad, Atera: SF-SWU; 2001.

[19] WHO. Chemical releases associated with earthquakes, Available from: https://www.who.int/publications/i/item/chemical-releases-associated-with-earthquakes

[20] Data world. US DOE/NNSA Response to 2011 Fukushima Incident: March 2011 Aerial Data, Available from: https://data.world/us-doe-gov/46ef4d73-da7c-4f15-9648-154344708268

[21] Data world. Nepal - Causalities caused by earthquake, 2015, Available from: https://data.world/opennepal/d63ec0b5-2e99-4991-b2c2-97b41eecec6c

[22] WHO: The response to the 2015 Nepal earthquakes: the value of preparedness, Available from: https://youtu.be/P29RknVelNM

[23] WNA. World Nuclear Association. Chernobyl Accident 1986, Available from: https://www.world-nuclear.org/information-library/safety-and-security/safety-of-plants/chernobyl-accident.aspx

[24] Health Effects of the Chernobyl Accident and Special Health Care Programme, Report of the UN Chernobyl Forum, Expert Group "Health", World Health Organization, 2006 (ISBN: 9789241594172)

[25] WNA. World Nuclear Association. Fukushima Daiichi Accident, Available from:https://www.world-nuclear.org/information-library/safety-and-security/safety-of-plants/fukushima-daiichi-accident.aspx

[26] IAEA. Fukushima Nuclear Accident

[27] Dimitrova D., Emergency conditions in disastrous situations. Shock.-In: Medicine of the disastrous situations. 1st ed. ARSO, S.; 2011, 374-383

Section 4

Mexico Early Waring System

Chapter 5

The Risk of Tsunamis in Mexico

Jaime Santos-Reyes and Tatiana Gouzeva

Abstract

This paper reviews the risk of tsunami in Mexico. It is highlighted that the Pacific coast of the country forms part of the so called "Ring of fire." Overall, the risk of tsunami that has the potentiality to affect communities along the Pacific coast is twofold: (a) local tsunami; that is, those triggered by earthquakes originating from the "Cocos," "Rivera," and the "North American" plates (high risk) and (b) the remote tsunamis, those generated elsewhere (e.g., Alaska, Japan, Chile) (low risk). Further, a preliminary model for "tsunami early warning" system for the case of Mexico is put forward.

Keywords: tsunami, earthquake, Mexico, tsunami early warning

1. Introduction

A *tsunami* has been defined as "a series of travelling waves of extremely long length and period, usually generated by disturbances associated with earthquake occurring below or near the ocean floor... Volcanic eruptions, submarine landslides, and coastal rock falls can also generate tsunamis, as can a large meteorite impacting the ocean" [1]. Also, tsunamis may be regarded as low frequency events but with high impacts in terms of human/infrastructure/economic losses. Their power of destruction has been more than evident in recent years [2–11]. It is believed that from the time period between 1998 and 2017, the losses inflicted by tsunami disasters were a total of US$280 billion and 251,770 causalities, in damages [7]. Moreover, the authors argue that the impact from this period has been 100 times higher than during the time period 1978–1997.

Following the 2004 tsunami in the Indian Ocean, there has been a large amount of literature published on several topics associated with tsunami science. For example, research has been conducted on the physics of tsunami waves [12], tsunami's impact and characteristics [1–3, 11, 13], tsunami early warning systems [14, 15], tsunami risk assessment [8, 10, 11, 16], geology's perspective [17–19], to mention a few.

Recent tsunamis have highlighted the need for an effective early warning system. An early warning is defined as "the provision of timely and effective information, through identified institutions, that allows individuals exposed to a hazard to take action to avoid or reduce their risk and prepare for effective response" [20]. Moreover, the United Nations Inter-Agency Secretariat of the International Strategy for Disaster Reduction (UN/ISDR) argues that an "effective early warning system" should include the following four key elements: "the knowledge of risks," "the technical monitoring and warning service," "dissemination & communication of meaningful warnings to those at risk," and "the public awareness and preparedness to react to warnings" [20, 21].

The objective of the paper is to highlight the tsunami risk in Mexico. The data showed in the paper are based on previous studies on tsunamis in the country [15, 22]. Further, a preliminary "tsunami early warning" system which aims at integrating, for example, the four key elements proposed by the UNISDR [20] for the case of Mexico is presented.

2. The risk of tsunamis in Mexico

The "Pacific ring of fire" belt covers a vast area of highly active tectonic plate boundaries where most of the earthquakes originate and active volcanoes (**Figure 1**). It is believed that the quarts of all the volcanoes in the world are in the ring [23].

Moreover, the Ring of fire runs through several countries, such as Canada, USA, Russia, Chile, Peru, Guatemala, New Zealand, Japan, Indonesia, Philippines, and Mexico.

Regarding the tsunami risk in Mexico, studies based on tsunami historical data showed that there are two zones of tsunami threat: local (i.e., generation of tsunamis) and remote (i.e., arrival of tsunamis) (**Figure 2**) [15, 22]. The authors defined these two zones by considering the nature of the faulting and tectonic plate interaction. In the subsequent subsection, each of these will be addressed.

2.1 Local tsunami risk

According to [15, 22] at the west of the "Rivera plate" and along the "Middle America trench," the "Cocos plate" subduction beneath the "North American plate" at rates of 2.5–7.7 cm/year (**Figure 2**). Given the fact, that large earthquakes occur in this region; therefore, the zone has been regarded as a generator of tsunamis (**Table 1** and **Figure 3**).

According to the historical data, the generated tsunamis that produced the highest wave heights were those that occurred in 1925 (7–11 m), 1932 (9–10 m), 1995 (2.9–5.10 m), and 1985 (1–3 m). For example, the 1985 earthquake of 8.1 Ms of

Figure 1.
The "Ring of fire" [23].

Figure 2.
Mexico's local and remote tsunami threat [15, 22].

Year	Region	Magnitude	Tsunami (places hit, Mexico)	Max. height waves (m)
1732	Guerrero	—	Acapulco	4.0
1754	Guerrero	—	Acapulco	5.0
1787	Guerrero	>8.0	Acapulco	3–8
1787	Oaxaca	—	Juquila	4.0
			Pochutla	4.0
1820	Guerrero	7.6	Acapulco	4.0
1852	B. C.	—	Río Colorado	3.0
1907	Guerrero	7.6	Acapulco	2.0
1925	Guerrero	7.0	Zihuatanejo	7.0–11.0
1932	Jalisco	8.2	Manzanillo	2.0
			San Pedrito	3.0
1932	Jalisco	7.8	Manzanillo	1.0
1932	Jalisco	6.9	Cuyutlán	9.0–10.0
1948	Nayarit	6.9	Islas Marias	2.0–5.0
1957	Guerrero	7.8	Acapulco	2.6
1973	Colima	7.6	Manzanillo	1.1
1978	Oaxaca	7.6	Puerto Escondido	1.5
1979	Guerrero		Acapulco	1.3
1985	Michoacán	8.0	Lázaro Cárdenas	2.5
			Ixtapa Zihuatanejo	3.0
			Playa Azul	2.5
			Acapulco	1.1
			Manzanillo	1.0
1985	Michoacan	7.8	Acapulco	1.2
			Zihuatanejo	2.5

Year	Region	Magnitude	Tsunami (places hit, Mexico)	Max. height waves (m)
1995	Colima	8.1	Boca de Iguanas	5.10
			Barra de Navidad	5.10
			San Mateo	4.90
			Melaque	4.50
			Cuastecomate	4.40
			El Tecuán	3.80
			Punta Careyes	3.50
			Chamela	3.20
			Pérula	3.40
			Punta ⌣ alaca⌣ ec	2.90
2003	Colima	7.8	Man anillo	1.22
2017	Ch⌣ pa⌣	8.1	Sali⌣ Cruz	1.10

Table 1.
Local tsunamis-only those with height >1.0 m is shown [22].

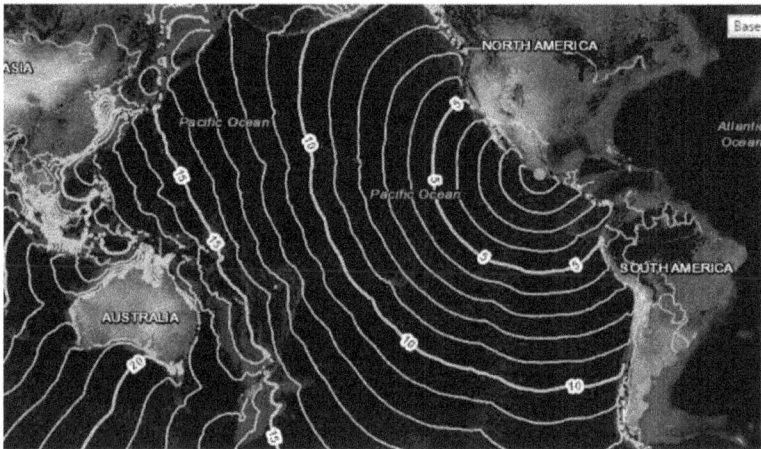

Figure 3.
Local tsunamis in the pacific coast of Mexico [24].

magnitude generated a tsunami that affected several communities in this zone. It is believed that a key infrastructure port was affected with waves of 2.5 m and flooded the area about 500 m inland [15]. Also, several tourist resorts were affected by the tsunami; for example, waves for up to 2.5 m high were observed in Playa de Azul [15].

Interestingly, a day after the main earthquake, a 7.5 Ms aftershock hit the zone; it is thought the generated tsunami affected a local fishing community with waves ranging from 2 to 3 m high [15].

2.2 Remote tsunami risk

It is belie ved that on he No th es. of he "Riv ra pl te" (Figur 2), along the Gulf of Cali ornia w ere th Pa ii c Plate slides no rth w th espect o th North American plate, gene ation of tsunamis in this zone is unlikely [15, 22]. This is consistent with historical data (**Table 2**); it can be seen that data on "small" and "moderate" tsunamis generated by remote sources; for example, the two most recent 2010 Chile and the 2011 tsunamis (**Figure 4**) were the maximum wave heights registered were <1.0 m.

Date	Region	Magnitude	Tsunami (places hit, Mexico)	Max. height waves (m)
1952	Kamchatka, USSR	8.3	La Paz, BCS	0.5
			Salina Cruz	1.2
1957	Aleutian Islands	8.3	Ensenada, B.C.	1.0
1960	Chile	8.5	Ensenada, B.C.	2.5
			La Paz, B.C.S.	1.5
			Mazatlán	1.1
			Acapulco	1.9
			Salina Cruz	1.6
1960	Peru	8	Acapulco	0.10
1963	Kuril, Japan, USSR	8.	Acapulco	<1.0
			Salina Cruz	
			Mazatlan	
			La Paz, B.C.S.	
1964	Alaska	8.4	Ensenada, B.C.	2.4
			Manzanillo	1.2
			Acapulco	1.1
			Salina Cruz	0.8
1968	Japan	8.0	Ensenada, B.C.	<1.0
			Manzanillo	
			Acapulco	
1975	Hawaii	7.2	Ensenada, B.C.	<1.0
			Manzanillo	
			Puerto Vallarta	
			Acapulco	
1976	Kermadec Islands	7.3	San Lucas, B.C.S.	<1.0
			Puerto Vallarta	
			Manzanillo	
			Acapulco	
1995	Chile	7.8	Cabo San Lucas	<1.0
2004	Indonesia	9.0	Manzanillo	1.22
			Lazaro Cardenas	0.24
			Zihuatanejo	0.60
2010	Chile	8.8–9.0	Manzanillo	0.32
			Cabo San Lucas	0.36
			Acapulco	0.62
2011	Japan	9.0	Ensenada, B.C.	0.70
			Huatulco	0.70
			Puerto Angel	0.29
			Acapulco	0.72
2018	Indonesia	7.5	—	—
2018	Indonesia	AK Vulcano tsunami	—	—

Table 2.
Remote tsunami -historical data taken from [22] with the exception of the last two tsunamis that occurred in 2018.

However, it is worth mentioning that the historical data showed that there were two tsunamis that registered the height of waves up to 2.4 and 2.5 m; that is, those generated in Chile (1960) and Alaska (1964), respectively (**Table 2**).

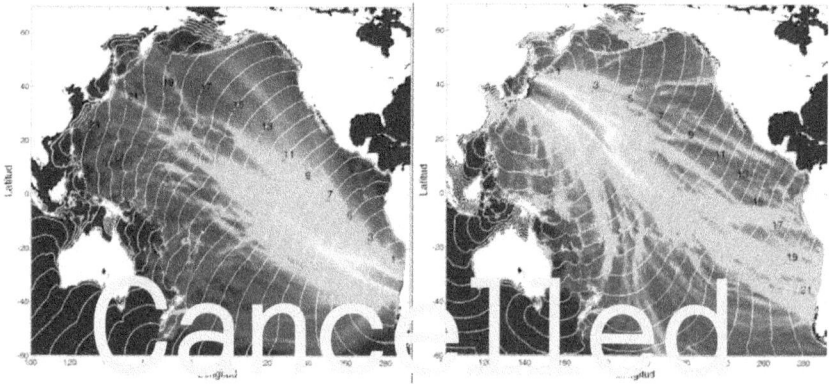

Figure 4.
The 2010 Chile tsunami (left) and the 2011 tsunami in Japan (right) [25].

3. A Mexican tsunami early warning system

The previous section and the most recent tsunami events [2–4] have highlighted the need for an effective tsunami early warning system (TEWS). A system should include "tsunami early warning coordination centres (TEWCC)" covering the whole of the Pacific coast of Mexico. Moreover, the system should also include earthquake early warning (EEW) systems. Furthermore, these systems should be explicitly "people-centered" [21, 26]. However, only those aspects associated with the features of a TEWS will be discussed in some detail. The proposed model is based on previous research on issues related to safety and disaster management systems [27–29].

Figure 5 shows what is called a "structural organization" of the model, which comprises essentially a set of five highly interrelated subsystems (systems 1–5).

In the context of this case study, the overall function of systems 2–5 (MTEW-SMU) is to establish the key tsunami safety policies aiming at maintaining tsunami risk within an acceptable range; this implies allocating the necessary resources, for example, to build response capabilities at national and community levels. System 1, on the other hand, embraces the TNZO (Tsunami Northern Zone Operations) and TSZO (Tsunami Southern Zone Operations) with their associated management units (TNZ-SMU and TSZ-SMU). These two operations of system 1 were considered given the fact that the risk of tsunamis comes from local and remote tsunami sources as mentioned in Section 2. See Section 2.

It is important to highlight that one of the key functions within the MTEW-SMU is that related to system 2, which is associated with what it is called here MTEW-CC (Mexican Tsunami Early Warning Coordination Centre); its key function is the monitoring of the TSZ-CC (Tsunami Southern Zone-Coordination Centre) and TNZ-CC (Tsunami Northern Zone Coordination Centre), as shown in **Figure 5**. The process of the flow of key information and decision making is described in **Table 3**.

Following the 2014 tsunami in the Indian Ocean, the need for a tsunami warning system (TWS) was more than evident; however, it may be argued that the existing TWS may be deficient in dealing with the mitigation of impacts of such events; moreover, there are still regions worldwide without such systems (**Table 4**).

Recent tsunami disasters have highlighted some of these deficiencies; for example, in the case of the 2010 tsunami in Chile, the entity in charge of issuing a tsunami warning failed to do so [24] (see "action point" "2"and "7" in **Figure 5**

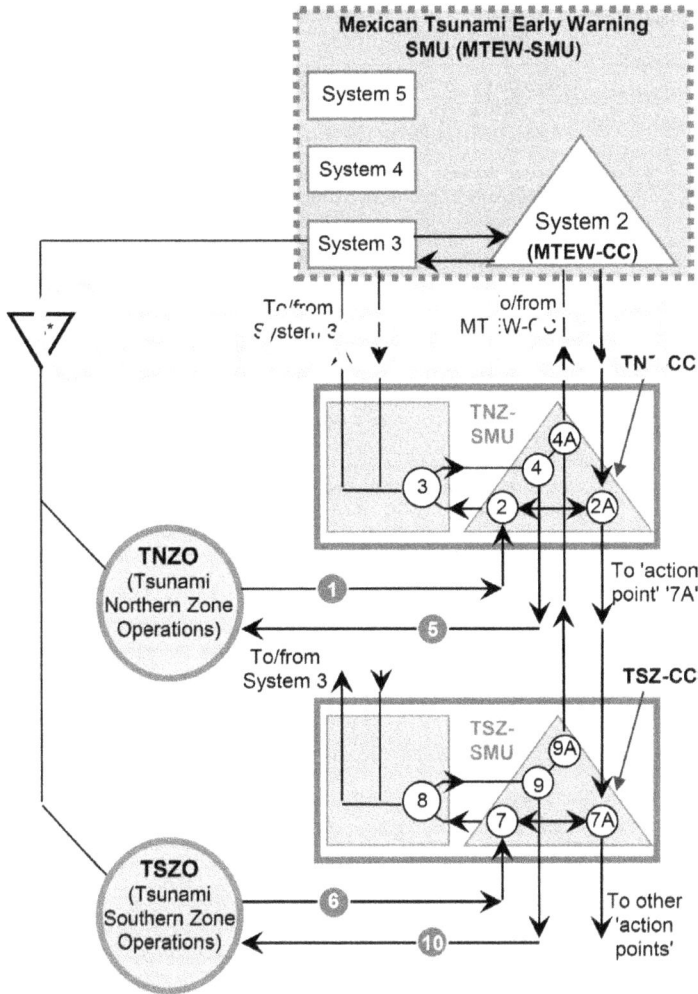

Figure 5.
A Mexican tsunami early warning system (MTEWS). Source: Tables 3 and 4 present details of the acronyms and action points.

"Action-points" (Figure 2)	Description
"1"	Data on key variables monitored by the TNZ-CC (pressure sensors, tide gauges, etc.) It should also be mentioned that this information is provided by the PTWC (Pacific Tsunami Warning Centre) [30].
"2" and "2A"	In "2," the tsunami risk is assessed, if k~~ ~~ariab'~ not within the accept~ble criteria (e.g., a tsunami) then it issues the warning of tsunami to "2A" which in tu 1 issues the warni g t the SZ CC, e er f the isk i low (Se tion .)
"3" and "4"	ti s p ints "3 nd 4" p n dd "p~ T rest res ond th tsunami warning, for example, design of risk maps, plans to conduct drills, plans to warn and evacuate the vulnerable communities within TNZO. Action point "3" also issues the tsunami warning to MTEW-SMU, that is, to system 3.
"4A"	It communicates the measures taken to the response of the tsunami to the MTEW-CC, where it may devise further actions given its synergistic view of the total system through system 3, as shown in **Figure 5**.

"Action-points" (Figure 2)	Description
"5"	It issues the tsunami warning to the affected communities within this zone (e.g., B.C, B.C.S., Sinaloa, Manzanillo, etc.). It implements all the measures to mitigate the impact of the tsunami risk, for example, evacuation to safe areas, etc.
"6"	In contrast with the case of action point "1," the monitoring system in place should assess all the data being received from the key variables in real time, given the fact that a large earthquake could trigger a local tsunami (Section 2.1).
"7" and "7A"	Data analysis, if the risk of a tsunami is being detected, then it issues the warning action points "8" and "7A." B... aivi... ... info ...ion, ... n poir "7A" c mmur'...s it t 'T' Z-CC" through a... oin "2A
"8" and "9"	I ...Le..e w.ere .he p.....ng o. ..ective ....easu...s aimi... .. ...itigating the impact of the tsunami within the TSZO; for example, educating communities on what actions to take in case of a tsunami, design of risk maps, plans for drillings, etc. Action point "8" also issues the tsunami warning to MTEW-SMU, that is, to system 3.
"9A"	As with the case of "4A," it communicates the measures taken to the response of the tsunami to the MTEW-CC, where it may devise further actions given its synergistic view of the total system through system 3, as shown in **Figure 5**.
"10"	Similarly, as in "5," it issues the tsunami warning to the affected communities within this zone (e.g. Acapulco, Oaxaca, Manzanillo, Zihuatanejo, etc.). Most importantly, it implements all the necessary measures to mitigate the impact of the tsunami, for example, evacuation to safe areas, etc. Moreover, it also implements plans to relocate the affected people to safe areas if necessary.

Table 3.
Description of the key actions points of the model in **Figure 3.**

System	Acronym	Example-SMU	Example-operations
Systems 2–5	MTEW-SMU ("Mexican Tsunami Early Warning-SMU") MTEW-CC ("Mexican Tsunami Early Warning Coordination Centre")	Head of the Mexican Navy, Civil protection personnel, etc.	Control centers
System 1	TNZ-SMU ("Tsunami Northern Zone- SMU") TNZO ("Tsunami Northern Zone Operations") TNZ-CC ("Tsunami Northern Zone- Coordination Centre)	Control centers Control centers	Local infrastructure (gauges, pressure sensors, etc.) Local communities in the zone
System 1	TSZ-SMU ("Tsunami Southern Zone-SMU") TSZO ("Tsunami Southern Zone Operations") TNZ-CC ("tsunami Southern Zone- Coordination Centre)	Control centers Control centers	Local infrastructure (gauges, pressure sensors, etc.) Local communities in the zone

Table 4.
Examples of the ' .y p. vers who/what perform some of the funct. ns of t e systems of the mod '.

and **Table 3**, Th' f. lur tc pe1 ʼo. n tʲ is ·cti· n cc ntrib ιteɑ ʼo f ίa. 'tieʳ in the coastal communities. More recently, the 28 September Sulawesi tsunami and the 24 December Anak Krakatau (AK) volcano tsunami, both in Indonesia, illustrate deficiencies in TWS too. In the former case, the tsunami warning was issued but the warning was lifted over 30 minutes [4]. However, the city of Palu, located in a narrow bay, was hit hard with waves reaching 6 m of height; why were not they warned? The head of the BMKG (Indonesia Agency for Meteorology, Climatology and Geophysics) argued that "we have no observation data at Palu…," "If we had a tide gauge or proper data in Palu, of course it would have been better" [4]. The tsunami (and earthquake) killed over 2000 people [2]. Finally, regarding the AK volcano tsunami, it is thought that there was not a tsunami warning system for the case of volcano-induced tsunamis, the tsunami killed 437 people [3].

It may be argued that a TWS should not be only concerned with the technical aspects (e.g., tidal gauge, network of buoys, etc.), but also the organizational and human components. In other words, there is a need for an effective tsunami early warning system that is able to consider all these components in a coherent manner, such as the system being proposed in Peru and elsewhere. Moreover these systems should be "people centered" [2, 6].

4. Conclusions

This paper has presented the risk of tsunamis in Mexico. The approach has been a review of existing literature on historical data of tsunami occurrence in Mexico. The literature survey showed that the tsunami threat comes from local and remote zones. Overall, the review showed that the highest tsunami risk comes from tsunamis induced by earthquakes occurring in the Southern zone of the country (i.e., local zone). The paper has also put forward a preliminary model of a TEWS (Tsunami Early Warning System) for the case of Mexico. However, it needs further research to design the whole networks of the flows of information not only for the case of tsunamis, but also for the case of earthquake early warning system.

Acknowledgements

This research was supported by the following grants: CONACYT-No:248219; SIP-IPN-20201790.

Conflict of interest

The authors declare that they have no competing interests.

Author details

Jaime Santos-Reyes[1*] and Tatiana Gouzeva[2]

1 Grupo de investigación SARACS, ESIME, ZAC., Instituto Politécnico Nacional, Mexico

2 Grupo de investigación SARACS, Instituto Politécnico Nacional, Mexico

*Address all correspondence to: jrsantos@hotmail.com

IntechOpen

References

[1] UNESCO/IOC (United Nations Educational, Scientific and Cultural Organization/Intergovernmental Oceanographic Commission). 2013. Tsunami Glossary. Revised Edition 2013. IOC Technical series, 85. UNESCO, Paris (IOC/2008/TS/85rev)

[2] Widiyan o W, Sar wo o F ?, Hsiao SC, I. anar a a . Po t-ev er field survey o 2o Sep mber Sulawesi earthquake and tsunami. Natural Hazards and Earth System Sciences. 2019;**19**:2781-2794

[3] Muhari A, Heidarzadeh M, Susmoro H, Nugroho HD, Kriswati E, Supartoyo Wijanarto AB, et al. The December 2018 Anak Krakatau Volcano Tsunami as inferred from post-tsunami field survey and spectral analysis. Pure and Applied Geophysics. 2019;**176**:5219-5233

[4] BBC. Indonesia earthquake and the tsunami: How warning system failed the victims. 2018. Available at: https://www.bbc.com/news/world-asia-45663054 [Accessed: 24 December 2019]

[5] Valenzuela IB, Camus PM. Chile pre y post catástrofe: Algunas claves para aproximarse a los desafíos de reconstrucción. In: Brain V, Mora C, editors. Emergencia y reconstruccion: El antes y después del terremoto y tsunami del 27 F en Chile. Chile: Fundacion Mapfre y Pontificia Universidad Católica de Chile; 1991

[6] Fakhrul-Razi A, Ridwan-Wong MM, Mat-Said A. Consequences of the 2004 Indian Ocea su ami in Malaysia. Safety Scie ce. 2020,**12** 61 -6. 1

[7] Imamura F, Penmellen-Boret S, Suppasri A, Muhari A. Recent occurrences of serious tsunami damage and the future challenges of tsunami disaster risk reduction. Progress in Disaster Science. 2019;**1**:100009

[8] Moreno J, Lara A, Torres M. Community resilience in response to the 2010 tsunami in Chile: The survival of a small-scale fishing community. International Journal of Disaster Risk Reduction. 2019;**33**:376-384

[9] L iba 7 McCullough H Mungov G, rne J, St ker . oh r arthquake and ts nam da avai ble f om the ational ceanic and Atmospheric Administration/National Geophysical Data Center, Geomatics. Natural Hazards and Risk. 2011;**2**(4):305-323. DOI: 10.1080/19475705.2011.632443

[10] Leon-Canales J, Vicuña del Río M, Gubler A. Increasing tsunami risk through intensive urban densification in metropolitan areas: A longitudinal analysis in Viña del Mar, Chile. International Journal of Disaster Risk Reduction. 2019;**41**:101312

[11] Li Z, Yu H, Chen X, Zhang G, Ma D. Tsunami-induced traffic evacuation strategy optimization. Transportation Research Part D. 2019;**77**:535-559

[12] Robke BR, Vott A. The tsunami phenomenon. Progress in Oceanography. 2017;**159**:296-322

[13] Goff J, Terry JP, Chagué-Goff C, Goto K. What is a mega-tsunami? Marine Geology. 2014;**358**:12-17

[14] Chaturvedi SK. A case study of tsunami detection system and ocean wave imaging mechanism using radar. Journal of Ocean Engineering and Science. 2019;**4**:203-210

[1] Fe rara SF, anc he Aj. The ts na i the eat n the Mex an west coast: A historical analysis and recommendations for hazard mitigation. Natural Hazards. 1991;**4**:301-316

[16] Jaimes MA, Reinoso E, Ordaz M, Huerta B, Silva R, Mendoza E, et al.

A new approach to probabilistic earthquake-induced tsunami risk assessment. Ocean and Coastal Management. 2016;**119**:68-75

[17] Trejo-Gómez E, Ortiz M, Nuñez-Cornú FJ. Source model of the Octobre 9, 1995 Jalisco-Colima Tsunami as constrained by field survey reports, and on the numerical simulation of the tsunami. Geofisica Internacional. 2015;**54**(2):249-159

[18] Roy PD, Jonathan MP, Consuelo-Macias M, Sanchez JL, Lozano R, Srinivasalu S. Geological characteristics of 2011 Japan tsunami sediments deposited along the coast of southwestern Mexico. Chemie der Erde. 2012;**72**:91-95

[19] Ramirez-Herrera MT, Bogalo MF, Cerny J, Goguitchaichvili A, Corona N, Machain ML, et al. Historic and ancient tsunamis uncovered on the Jalisco-Colima Pacific coast, the Mexican subduction zone. Geomorphology. 2016;**259**:90-104

[20] United Nations International Strategy for Disaster Reduction (UN/ISDR). Global Survey of Early Warning Systems. Geneva, Switzerland: UNISDR; 2006

[21] Bashir R. Global early warning systems for natural hazards: Systematic and people centred. Philosophical Transactions of the Royal Society A. 2006;**364**:2167-2182

[22] Ferraras SF, Dominguez-Mora R, Gutierrez-Martinez CA. Tsunamis. Mexico: Centro Nacional de Prevencion de Desastres (CENAPRED), Secretaria de Gobernación; 2014

[23] USSG. Ring of fire map. Available at: http://www.geologyin. com/2018/01/the-ring-of-fire. html#kMGfZ4P00Ue21LoI.99 [Accessed: 24 December 2019]

[24] NOAA. Estimated tsunami travel times to coastal locations-Acapulco Mexico. National Centers for Environmental Information (NOAA). 2019. Available at: https://maps.ngdc. noaa.gov/viewers/ttt_coastal_locations/ [Accessed: 24 December 2019]

[25] Ortiz-Huerta LG, Ortiz-Figueroa M. Cómo me puedo preparar ante un tsunami. In: Guía para vivir nuestro. Mexico: SEGOB, CENAPRED, CICESE; 2014

[26] Santos-Reyes J. How useful are earthquake early warnings? The case of the 2017 earthquakes in Mexico City. International Journal of Disaster Risk Reduction. 2019;**40**(101148):1-11

[27] Santos-Reyes J, Padilla-Perez D, Beard AN. Transport infrastructure interdependency: Metro's failure propagation in the road transport system in Mexico City. Sustainability. 2019;**11**(4757):1-24. DOI: 10.3390/su11174757

[28] Santos-Reyes J, Beard AN. An analysis of the 1996 channel tunnel fire. Proceedings of the Institution of Mechanical Engineers, Part F. 2017;**231**(8):850-876. DOI: 10.1177/0954409716647093

[29] Santos-Reyes J, Beard AN. Applying the SDMS model to the analysis of the Tabasco flood disaster in Mexico. Human and Ecological Risk Assessment: An International Journal. 2011;**17**(3):646-677. DOI: 10.1080/10807039.2011.571099

[30] PTWC. Pacific Tsunamy Warning Centre. 2019. Available at: https://ptwc. weather.gov [Accessed: 24 December 2019]

www.ingramcontent.com/pod-product-compliance
Lightning Source LLC
Chambersburg PA
CBHW081234190326
41458CB00016B/5779